Table of Contents

INTRODUCTION

LECTURE GUIDES

Table of Contents

The Secrets of Mental Math

Arthur T. Benjamin, Ph.D.

PUBLISHED BY:

THE GREAT COURSES
Corporate Headquarters
4840 Westfields Boulevard, Suite 500
Chantilly, Virginia 20151-2299
Phone: 1-800-832-2412
Fax: 703-378-3819
www.thegreatcourses.com

Printed in the United States of America

Arthur T. Benjamin, Ph.D.

Professor of Mathematics
Harvey Mudd College

Professor Arthur T. Benjamin is a Professor of Mathematics at Harvey Mudd College. He graduated from Carnegie Mellon University in 1983, where he earned a B.S. in Applied Mathematics with university honors. He received his Ph.D. in Mathematical Sciences in 1989 from Johns Hopkins University, where he was supported by a National Science Foundation graduate fellowship and a Rufus P. Isaacs fellowship. Since 1989, Professor Benjamin has been a faculty member of the Mathematics Department at Harvey Mudd College, where he has served as department chair. He has spent sabbatical visits at Caltech, Brandeis University, and the University of New South Wales in Sydney, Australia.

In 1999, Professor Benjamin received the Southern California Section of the Mathematical Association of America (MAA) Award for Distinguished College or University Teaching of Mathematics, and in 2000, he received the MAA Deborah and Franklin Tepper Haimo National Award for Distinguished College or University Teaching of Mathematics. He was also named the 2006–2008 George Pólya Lecturer by the MAA.

Professor Benjamin's research interests include combinatorics, game theory, and number theory, with a special fondness for Fibonacci numbers. Many of these ideas appear in his book (coauthored with Jennifer Quinn) *Proofs That Really Count: The Art of Combinatorial Proof*, published by the MAA. In 2006, that book received the MAA's Beckenbach Book Prize. From 2004 to 2008, Professors Benjamin and Quinn served as the coeditors of *Math Horizons* magazine, which is published by the MAA and enjoyed by more than 20,000 readers, mostly undergraduate math students and their teachers. In 2009, the MAA published Professor Benjamin's latest book, *Biscuits of Number Theory*, coedited with Ezra Brown.

Professor Benjamin is also a professional magician. He has given more than 1000 "mathemagics" shows to audiences all over the world (from primary schools to scientific conferences), in which he demonstrates and explains his calculating talents. His techniques are explained in his book *Secrets of Mental Math: The Mathemagician's Guide to Lightning Calculation and Amazing Math Tricks*. Prolific math and science writer Martin Gardner calls it "the clearest, simplest, most entertaining, and best book yet on the art of calculating in your head." An avid game player, Professor Benjamin was winner of the American Backgammon Tour in 1997.

Professor Benjamin has appeared on dozens of television and radio programs, including the *Today* show, *The Colbert Report*, CNN, and National Public Radio. He has been featured in *Scientific American*, *Omni*, *Discover*, *People*, *Esquire*, *The New York Times*, the *Los Angeles Times*, and *Reader's Digest*. In 2005, *Reader's Digest* called him "America's Best Math Whiz." ■

The Secrets of Mental Math

Scope:

Most of the mathematics that we learn in school is taught to us on paper with the expectation that we will solve problems on paper. But there is joy and lifelong value in being able to do mathematics in your head. In school, learning how to do math in your head quickly and accurately can be empowering. In this course, you will learn to solve many problems using multiple strategies that reinforce number sense, which can be helpful in all mathematics courses. Success at doing mental calculation and estimation can also lead to improvement on several standardized tests.

We encounter numbers on a daily basis outside of school, including many situations in which it is just not practical to pull out a calculator, from buying groceries to reading the newspaper to negotiating a car payment. And as we get older, research has shown that it is important to find activities that keep our minds active and sharp. Not only does mental math sharpen the mind, but it can also be a lot of fun.

Our first four lectures will focus on the nuts and bolts of mental math: addition, subtraction, multiplication, and division. Often, we will see that there is more than one way to solve a problem, and we will motivate many of the problems with real-world applications.

Once we have mastery of the basics of mental math, we will branch out in interesting directions. Lecture 5 offers techniques for easily finding approximate answers when we don't need complete accuracy. Lecture 6 is devoted to pencil-and-paper mathematics but done in ways that are seldom taught in school; we'll see that we can simply write down the answer to a multiplication, division, or square root problem without any intermediate results. This lecture also shows some interesting ways to verify an answer's correctness. In Lecture 7, we go beyond the basics to explore advanced multiplication techniques that allow many large multiplication problems to be dramatically simplified.

In Lecture 8, we explore long division, short division, and Vedic division, a fascinating technique that can be used to generate answers faster than any method you may have seen before. Lecture 9 will teach you how to improve your memory for numbers using a phonetic code. Applying this code allows us to perform even larger mental calculations, but it can also be used for memorizing dates, phone numbers, and your favorite mathematical constants. Speaking of dates, one of my favorite feats of mental calculation is being able to determine the day of the week of any date in history. This is actually a very useful skill to possess. It's not every day that someone asks you for the square root of a number, but you probably encounter dates every day of your life, and it is quite convenient to be able to figure out days of the week. You will learn how to do this in Lecture 10.

In Lecture 11, we venture into the world of advanced multiplication; here, we'll see how to square 3- and 4-digit numbers, find approximate cubes of 2-digit numbers, and multiply 2- and 3-digit numbers together. In our final lecture, you will learn how to do enormous calculations, such as multiplying two 5-digit numbers, and discuss the techniques used by other world-record lightning calculators. Even if you do not aspire to be a grandmaster mathemagician, you will still benefit tremendously by acquiring the skills taught in this course. ■

Acknowledgments

Putting this course together has been extremely gratifying, and there are several people I wish to thank. It has been a pleasure working with the very professional staff of The Great Courses, including Lucinda Robb, Marcy MacDonald, Zachary Rhoades, and especially Jay Tate. Thanks to Professor Stephen Lucas, who provided me with valuable historical information, and to calculating protégés Ethan Brown and Adam Varney for proof-watching this course. Several groups gave me the opportunity to practice these lectures for live audiences, who provided valuable feedback. In particular, I am grateful to the North Dakota Department of Public Instruction, Professor Sarah Rundell of Dennison University, Dr. Daniel Doak of Ohio Valley University, and Lisa Loop of the Claremont Graduate University Teacher Education Program.

Finally, I wish to thank my daughters, Laurel and Ariel, for their patience and understanding and, most of all, my wife, Deena, for all her assistance and support during this project.

Arthur Benjamin

Claremont, California

Math in Your Head!

Lecture 1

Just by watching this course, you will learn all the techniques that are required to become a fast mental calculator, but if you want to actually improve your calculating abilities, then just like with any skill, you need to practice.

In school, most of the math we learn is done with pencil and paper, yet in many situations, it makes more sense to do problems in your head. The ability to do rapid mental calculation can help students achieve higher scores on standardized tests and can keep the mind sharp as we age.

One of the first mental math tips you can practice is to calculate from **left to right**, rather than **right to left**. On paper, you might add 2300 + 45 from right to left, but in your head, it's more natural and faster to add from left to right.

These lectures assume that you know the multiplication table, but there are some tricks to memorizing it that may be of interest to parents and teachers. I teach students the multiples of 3, for example, by first having them practice counting by 3s, then giving them the multiplication problems in order (3×1, 3×2 …) so that they associate the problems with the counting sequence. Finally, I mix up the problems so that the students can practice them out of sequence.

The ability to do rapid mental calculation can help students achieve higher scores on standardized tests and can keep the mind sharp as we age.

There's also a simple trick to multiplying by 9s: The multiples of 9 have the property that their digits add up to 9 ($9 \times 2 = 18$ and $1 + 8 = 9$). Also, the first digit of the answer when multiplying by 9 is 1 less than the multiplier (e.g., $9 \times 3 = 27$ begins with 2).

In many ways, mental calculation is a process of simplification. For example, the problem 432 × 3 sounds hard, but it's the sum of three easy problems: 3 × 400 = 1200, 3 × 30 = 90, and 3 × 2 = 6; 1200 + 90 + 6 = 1296. Notice that when adding the numbers, it's easier to add from largest to smallest, rather than smallest to largest.

Again, doing mental calculations from left to right is also generally easier because that's the way we read numbers. Consider 54 × 7. On paper, you might start by multiplying 7 × 4 to get 28, but when doing the problem mentally, it's better to start with 7 × 50 (350) to get an estimate of the answer. To get the exact answer, add the product of 7 × 50 and the product of 7 × 4: 350 + 28 = 378.

Below are some additional techniques that you can start using right away:

- The product of 11 and any 2-digit number begins and ends with the two digits of the multiplier; the number in the middle is the sum of the original two digits. Example: 23 × 11 → 2 + 3 = 5; answer: 253. For a multiplier whose digits sum to a number greater than 9, you have to carry. Example: 85 × 11 → 8 + 5 = 13; carry the 1 from 13 to the 8; answer: 935.

- The product of 11 and any 3-digit number also begins and ends with the first and last digits of the multiplier, although the first digit can change from carries. In the middle, insert the result of adding the first and second digits and the second and third digits. Example: 314 × 11 → 3 + 1 = 4 and 1 + 4 = 5; answer: 3454.

- To square a 2-digit number that ends in 5, multiply the first digit in the number by the next higher digit, then attach 25 at the end. Example: 35^2 → 3 × 4 = 12; answer: 1225. For 3-digit numbers, multiply the first two numbers together by the next higher number, then attach 25. Example: 305^2 → 30 × 31 = 930; answer: 93,025.

- To multiply two 2-digit numbers that have the same first digits and last digits that sum to 10, multiply the first digit by the next higher digit, then attach the product of the last digits in the original two numbers. Example: $84 \times 86 \rightarrow 8 \times 9 = 72$ and $4 \times 6 = 24$; answer: 7224.

- To multiply a number between 10 and 20 by a 1-digit number, multiply the 1-digit number by 10, then multiply it by the second digit in the 2-digit number, and add the products. Example: $13 \times 6 \rightarrow (6 \times 10) + (6 \times 3) = 60 + 18$; answer: 78.

- To multiply two numbers that are both between 10 and 20, add the first number and the last digit of the second number, multiply the result by 10, then add that result to the product of the last digits in both numbers of the original problem. Example: $13 \times 14 \rightarrow 13 + 4 = 17$, $17 \times 10 = 170$, $3 \times 4 = 12$, $170 + 12 = 182$; answer: 182. ■

Important Terms

left to right: The "right" way to do mental math.

right to left: The "wrong" way to do mental math.

Suggested Reading

Benjamin and Shermer, *Secrets of Mental Math: The Mathemagician's Guide to Lightning Calculation and Amazing Math Tricks*, chapter 0.

Hope, Reys, and Reys, *Mental Math in the Middle Grades.*

Julius, *Rapid Math Tricks and Tips: 30 Days to Number Power.*

Ryan, *Everyday Math for Everyday Life: A Handbook for When It Just Doesn't Add Up.*

Problems

The following mental addition and multiplication problems can be done almost immediately, just by listening to the numbers from left to right.

1. $23 + 5$

2. $23 + 50$

3. $500 + 23$

4. $5000 + 23$

5. $67 + 8$

6. $67 + 80$

7. $67 + 800$

8. $67 + 8000$

9. $30 + 6$

10. $300 + 24$

11. $2000 + 25$

12. $40 + 9$

13. $700 + 84$

14. $140 + 4$

15. $2500 + 20$

16. $2300 + 58$

17. 13×10

18. 13×100

19. 13×1000

20. 243×10

21. 243×100

22. 243×1000

23. 243×1 million

24. Fill out the standard 10-by-10 multiplication table as quickly as you can. It's probably easiest to fill it out one row at a time by counting.

25. Create an 8-by-9 multiplication table in which the rows represent the numbers from 2 to 9 and the columns represent the numbers from 11 to 19. For an extra challenge, fill out the squares in random order.

26. Create the multiplication table in which the rows and columns represent the numbers from 11 to 19. For an extra challenge, fill out the rows in random order. Be sure to use the shortcuts we learned in this lecture, including those for multiplying by 11.

The following multiplication problems can be done just by listening to the answer from left to right.

27. 41×2

28. 62×3

29. 72×4

30. 52×8

31. 207×3

32. 402×9

33. 543×2

Do the following multiplication problems using the shortcuts from this lecture.

34. 21×11

35. 17×11

36. 54×11

37. 35×11

38. 66×11

39. 79×11

40. 37×11

41. 29×11

42. 48×11

43. 93×11

44. 98×11

45. 135×11

46. 261×11

47. 863×11

48. 789 × 11

49. Quickly write down the squares of all 2-digit numbers that end in 5.

50. Since you can quickly multiply numbers between 10 and 20, write down the squares of the numbers 105, 115, 125, … 195, 205.

51. Square 995.

52. Compute 1005^2.

Exploit the shortcut for multiplying 2-digit numbers that begin with the same digit and whose last digits sum to 10 to do the following problems.

53. 21 × 29

54. 22 × 28

55. 23 × 27

56. 24 × 26

57. 25 × 25

58. 61 × 69

59. 62 × 68

60. 63 × 67

61. 64 × 66

62. 65 × 65

Solutions for this lecture begin on page 82.

Mental Addition and Subtraction

Lecture 2

The bad news is that most 3-digit subtraction problems require some sort of borrowing. But the good news is that they can be turned into easy addition problems.

When doing mental addition, we work one digit at a time. To add a 1-digit number, just add the 1s digits ($52 + 4 \rightarrow 2 + 4 = 6$, so $52 + 4 = 56$). With 2-digit numbers, first add the 10s digits, then the 1s digits ($62 + 24 \rightarrow 62 + 20 = 82$ and $82 + 4 = 86$).

With 3-digit numbers, addition is easy when one or both numbers are multiples of 100 ($400 + 567 = 967$) or when both numbers are multiples of 10 ($450 + 320 \rightarrow 450 + 300 = 750$ and $750 + 20 = 770$). Adding in this way is useful if you're counting calories.

To add 3-digit numbers, first add the 100s, then the 10s, then the 1s. For $314 + 159$, first add $314 + 100 = 414$. The problem is now simpler, $414 + 59$; keep the 400 in mind and focus on $14 + 59$. Add $14 + 50 = 64$, then add 9 to get 73. The answer to the original problem is 473.

We could do $766 + 489$ by adding the 100s, 10s, and 1s digits, but each step would involve a carry. Another way to do the problem is to notice that $489 = 500 - 11$; we can add $766 + 500$, then subtract 11 (answer: 1255). Addition problems that involve carrying can often be turned into easy subtraction problems.

With mental subtraction, we also work one digit at a time from left to right. With $74 - 29$, first subtract $74 - 20 = 54$. We know the answer to $54 - 9$ will be 40-something, and $14 - 9 = 5$, so the answer is 45.

A subtraction problem that would normally involve borrowing can usually be turned into an easy addition problem with no carrying. For $121 - 57$, subtract 60, then add back 3: $121 - 60 = 61$ and $61 + 3 = 64$.

With 3-digit numbers, we again subtract the 100s, the 10s, then the 1s. For 846 – 225, first subtract 200: 846 – 200 = 646. Keep the 600 in mind, then do 46 – 25 by subtracting 20, then subtracting 5: 46 – 20 = 26 and 26 – 5 = 21. The answer is 621.

Three-digit subtraction problems can often be turned into easy addition problems. For 835 – 497, treat 497 as 500 – 3. Subtract 835 – 500, then add back 3: 835 – 500 = 335 and 335 + 3 = 338.

Understanding **complements** helps in doing difficult subtraction. The complement of 75 is 25 because 75 + 25 = 100. To find the complement of a 2-digit number, find the number that when added to the first digit will yield 9 and the number that when added to the second digit will yield 10. For 75, notice that 7 + 2 = 9 and 5 + 5 = 10. If the number ends in 0, such as 80, then the complement will also end in 0. In this case, find the number that when added to the first digit will yield 10 instead of 9; the complement of 80 is 20.

Understanding complements helps in doing difficult subtraction.

Knowing that, let's try 835 – 467. We first subtract 500 (835 – 500 = 335), but then we need to add back something. How far is 467 from 500, or how far is 67 from 100? Find the complement of 67 (33) and add it to 335: 335 + 33 = 368.

To find 3-digit complements, find the numbers that will yield 9, 9, 10 when added to each of the digits. For example, the complement of 234 is 766. Exception: If the original number ends in 0, so will its complement, and the 0 will be preceded by the 2-digit complement. For example, the complement of 670 will end in 0, preceded by the complement of 67, which is 33; the complement of 670 is 330.

Three-digit complements are used frequently in making change. If an item costs $6.75 and you pay with a $10 bill, the change you get will be the complement of 675, namely, 325, $3.25. The same strategy works with change from $100. What's the change for $23.58? For the complement of

2358, the digits must add to 9, 9, 9, and 10. The change would be \$76.42. When you hear an amount like \$23.58, think that the dollars add to 99 and the cents add to 100. With \$23.58, 23 + 76 = 99 and 58 + 42 = 100. When making change from \$20, the idea is essentially the same, but the dollars add to 19 and the cents add to 100.

As you practice mental addition and subtraction, remember to work one digit at a time and look for opportunities to use complements that turn hard addition problems into easy subtraction problems and vice versa. ■

Important Term

complement: The distance between a number and a convenient round number, typically, 100 or 1000. For example, the complement of 43 is 57 since 43 + 57 = 100.

Suggested Reading

Benjamin and Shermer, *Secrets of Mental Math: The Mathemagician's Guide to Lightning Calculation and Amazing Math Tricks*, chapter 1.

Julius, *More Rapid Math Tricks and Tips: 30 Days to Number Mastery.*

———, *Rapid Math Tricks and Tips: 30 Days to Number Power.*

Kelly, *Short-Cut Math.*

Problems

Because mental addition and subtraction are the building blocks to all mental calculations, plenty of practice exercises are provided. Solve the following mental addition problems by calculating from left to right. For an *added* challenge, look away from the numbers after reading the problem.

1. 52 + 7

2. 93 + 4

3. 38 + 9

4. 77 + 5

5. 96 + 7

6. 40 + 36

7. 60 + 54

8. 56 + 70

9. 48 + 60

10. 53 + 31

11. 24 + 65

12. 45 + 35

13. 56 + 37

14. 75 + 19

15. 85 + 55

16. 27 + 78

17. 74 + 53

18. 86 + 68

19. 72 + 83

Do these 2-digit addition problems in two ways; make sure the second way involves subtraction.

20. 68 + 97

21. 74 + 69

22. 28 + 59

23. 48 + 93

Try these 3-digit addition problems. The problems gradually become more difficult. For the harder problems, it may be helpful to say the problem out loud before starting the calculation.

24. 800 + 300

25. 675 + 200

26. 235 + 800

27. 630 + 120

28. 750 + 370

29. 470 + 510

30. 980 + 240

31. 330 + 890

32. 246 + 810

33. 960 + 326

34. 130 + 579

35. 325 + 625

36. 575 + 675

37. 123 + 456

38. 205 + 108

39. 745 + 134

40. 341 + 191

41. 560 + 803

42. 566 + 185

43. 764 + 637

Do the next few problems in two ways; make sure the second way uses subtraction.

44. 787 + 899

45. 339 + 989

46. 797 + 166

47. 474 + 970

Do the following subtraction problems from left to right.

48. 97 – 6

49. 38 – 7

50. 81 – 6

51. 54 – 7

52. 92 – 30

53. 76 – 15

54. 89 – 55

55. 98 – 24

Do these problems two different ways. For the second way, begin by subtracting too much.

56. 73 – 59

57. 86 – 68

58. 74 – 57

59. 62 – 44

Try these 3-digit subtraction problems, working from left to right.

60. 716 – 505

61. 987 – 654

62. 768 – 222

63. 645 – 231

64. 781 – 416

Determine the complements of the following numbers, that is, their distance from 100.

65. 28

66. 51

67. 34

68. 87

69. 65

70. 70

71. 19

72. 93

Use complements to solve these problems.

73. 822 – 593

74. 614 – 372

75. 932 – 766

76. 743 – 385

77. 928 – 262

78. 532 – 182

79. 611 – 345

80. 724 – 476

Determine the complements of these 3-digit numbers, that is, their distance from 1000.

81. 772

82. 695

83. 849

84. 710

85. 128

86. 974

87. 551

Use complements to determine the correct amount of change.

88. \$2.71 from \$10

89. \$8.28 from \$10

90. \$3.24 from \$10

91. \$54.93 from \$100

92. \$86.18 from \$100

93. \$14.36 from \$20

94. \$12.75 from \$20

95. \$31.41 from \$50

The following addition and subtraction problems arise while doing mental multiplication problems and are worth practicing before beginning Lecture 3.

96. 350 + 35

97. 720 + 54

98. 240 + 32

99. 560 + 56

100. 4900 + 210

101. 1200 + 420

102. 1620 + 48

103. 7200 + 540

104. 3240 + 36

105. 2800 + 350

106. 2150 + 56

107. 800 – 12

108. 3600 – 63

109. 5600 – 28

110. 6300 – 108

Solutions for this lecture begin on page 89.

Go Forth and Multiply

Lecture 3

You've now seen everything you need to know about doing 3-digit-by-1-digit multiplication. ... [T]he basic idea is always the same. We calculate from left to right, and add numbers as we go.

Once you've mastered the multiplication table up through 10, you can multiply any two 1-digit numbers together. The next step is to multiply 2- and 3-digit numbers by 1-digit numbers. As we'll see, these 2-by-1s and 3-by-1s are the essential building blocks to all mental multiplication problems. Once you've mastered those skills, you will be able to multiply any 2-digit numbers.

We know how to multiply 1-digit numbers by numbers below 20, so let's warm up by doing a few simple 2-by-1 problems. For example, try 53 × 6. We start by multiplying 6 × 50 to get 300, then keep that 300 in mind. We know the answer will not change to 400 because the next step is to add the result of a 1-by-1 problem: 6 × 3. A 1-by-1 problem can't get any larger than 9 × 9, which is less than 100. Since 6 × 3 = 18, the answer to our original problem, 53 × 6, is 318.

Here's an area problem: Find the area of a triangle with a height of 14 inches and a base of 59 inches. The formula here is 1/2(*bh*), so we have to calculate 1/2 × (59 × 14). The commutative law allows us to multiply numbers in any order, so we rearrange the problem to (1/2 × 14) × 59. Half of 14 is 7, leaving us with the simplified problem 7 × 59. We multiply 7 × 50 to get 350, then 7 × 9 to get 63; we then add 350 + 63 = to get 413 square inches in the triangle. Another way to do the same calculation is to treat 59 × 7 as (7 × 60) – (7 × 1): 7 × 60 = 420 and 7 × 1 = 7; 420 – 7 = 413. This approach turns a hard addition problem into an easy subtraction problem. When you're first practicing mental math, it's helpful to do such problems both ways; if you get the same answer both times, you can be pretty sure it's right.

The goal of mental math is to solve the problem without writing anything down. At first, it's helpful to be able to see the problem, but as you gain skill, allow yourself to see only half of the problem. Enter the problem on a calculator, but don't hit the equals button until you have an answer. This allows you to see one number but not the other.

The **distributive law** tells us that 3 × 87 is the same as (3 × 80) + (3 × 7), but here's a more intuitive way to think about this concept: Imagine we have three bags containing 87 marbles each. Obviously, we have 3 × 87 marbles. But suppose we know that in each bag, 80 of the marbles are blue and 7 are crimson. The total number of marbles is still 3 × 87, but we can also think of the total as 3 × 80 (the number of blue marbles) and 3 × 7 (the number of crimson marbles). Drawing a picture can also help in understanding the distributive law.

Most 2-digit numbers can be factored into smaller numbers, and we can often take advantage of this.

We now turn to multiplying 3-digit numbers by 1-digit numbers. Again, we begin with a few warm-up problems. For 324 × 7, we start with 7 × 300 to get 2100. Then we do 7 × 20, which is 140. We add the first two results to get 2240; then we do 7 × 4 to get 28 and add that to 2240. The answer is 2268. One of the virtues of working from left to right is that this method gives us an idea of the overall answer; working from right to left tells us only what the last number in the answer will be. Another good reason to work from left to right is that you can often say part of the answer while you're still calculating, which helps to boost your memory.

Once you've mastered 2-by-1 and 3-by-1 multiplication, you can actually do most 2-by-2 multiplication problems, using the **factoring method**. Most 2-digit numbers can be factored into smaller numbers, and we can often take advantage of this. Consider the problem 23 × 16. When you see 16, think of it as 8 × 2, which makes the problem 23 × (8 × 2). First, multiply by 8 (8 × 20 = 160 and 8 × 3 = 24; 160 + 24 = 184), then multiply 184 × 2 to get the answer to the original problem, 368. We could also do this problem by thinking of 16 as 2 × 8 or as 4 × 4.

For most 2-by-1 and 3-by-1 multiplication problems, we use the **addition method**, but sometimes it may be faster to use subtraction. By practicing these skills, you will be able to move on to multiplying most 2-digit numbers together. ■

Important Terms

addition method: A method for multiplying numbers by breaking the problem into sums of numbers. For example, $4 \times 17 = (4 \times 10) + (4 \times 7) = 40 + 28 = 68$, or $41 \times 17 = (40 \times 17) + (1 \times 17) = 680 + 17 = 697$.

distributive law: The rule of arithmetic that combines addition with multiplication, specifically $a \times (b + c) = (a \times b) + (a \times c)$.

factoring method: A method for multiplying numbers by factoring one of the numbers into smaller parts. For example, $35 \times 14 = 35 \times 2 \times 7 = 70 \times 7 = 490$.

Suggested Reading

Benjamin and Shermer, *Secrets of Mental Math: The Mathemagician's Guide to Lightning Calculation and Amazing Math Tricks*, chapter 2.

Julius, *More Rapid Math Tricks and Tips: 30 Days to Number Mastery.*

———, *Rapid Math Tricks and Tips: 30 Days to Number Power.*

Kelly, *Short-Cut Math.*

Problems

Because 2-by-1 and 3-by-1 multiplication problems are so important, an ample number of practice problems are provided. Calculate the following 2-by-1 multiplication problems in your head using the addition method.

1. 40×8

2. 42×8

3. 20×4

4. 28×4

5. 56×6

6. 47×5

7. 45×8

8. 26×4

9. 68×7

10. 79×9

11. 54×3

12. 73×2

13. 75×8

14. 67×6

15. 83×7

16. 74×6

17. 66×3

18. 83×9

19. 29×9

20. 46×7

Calculate the following 2-by-1 multiplication problems in your head using the addition method and the subtraction method.

21. 89×9

22. 79×7

23. 98×3

24. 97×6

25. 48×7

The following problems arise while squaring 2-digit numbers or multiplying numbers that are close together. They are essentially 2-by-1 problems with a 0 attached.

26. 20×16

27. 20×24

28. 20×25

29. 20×26

30. 20×28

31. 20×30

32. 30×28

33. 30×32

34. 40×32

35. 30×42

36. 40×48

37. 50×44

38. 60×52

39. 60×68

40. 60×69

41. 70×72

42. 70×78

43. 80×84

44. 80×87

45. 90×82

46. 90×96

Here are some more problems that arise in the first step of a 2-by-2 multiplication problem.

47. 30×23

48. 60×13

49. 50×68

50. 90×26

51. 90×47

52. 40×12

53. 80×41

54. 90×66

55. 40×73

Calculate the following 3-by-1 problems in your head.

56. 600×7

57. 402×2

58. 360×6

59. 360×7

60. 390×7

61. 711×6

62. 581×2

63. 161×2

64. 616×7

65. 679×5

66. 747×2

67. 539×8

68. 143×4

69. 261×8

70. 624×6

71. 864×2

72. 772×6

73. 345×6

74. 456×6

75. 476×4

76. 572×9

77. 667×3

When squaring 3-digit numbers, the first step is to essentially do a 3-by-1 multiplication problem like the ones below.

78. 404×400

79. 226×200

80. 422×400

81. 110×200

82. 518×500

83. 340×300

84. 650×600

85. 270×200

86. 706×800

87. 162×200

88. 454×500

89. 664×700

Use the factoring method to multiply these 2-digit numbers together by turning the original problem into a 2-by-1 problem, followed by a 2-by-1 or 3-by-1 problem.

90. 43×14

91. 64×15

92. 75×16

93. 54×24

94. 89×72

In poker, there are 2,598,960 ways to be dealt 5 cards (from 52 different cards, where order is not important). Calculate the following multiplication problems that arise through counting poker hands.

95. The number of hands that are straights (40 of which are straight flushes) is

$10 \times 4^5 = 4 \times 4 \times 4 \times 4 \times 4 \times 10 = ???$

96. The number of hands that are flushes is

$(4 \times 13 \times 12 \times 11 \times 10 \times 9)/120 = 13 \times 11 \times 4 \times 9 = ???$

97. The number of hands that are four-of-a-kind is $13 \times 48 = ???$

98. The number of hands that are full houses is $13 \times 12 \times 4 \times 6 = ???$

Solutions for this lecture begin on page 97.

Divide and Conquer

Lecture 4

When I was a kid, I remember doing lots of 1-digit division problems on a bowling league. If I had a score of 45 after three frames, I would divide 45 by 3 to get 15, and would think, "At this rate, I'm on pace to get a score of 150."

We begin by reviewing some tricks for determining when one number divides evenly into another, then move on to 1-digit division. Let's first try 79 ÷ 3. On paper, you might write 3 goes into 7 twice, subtract 6, then bring down the 9, and so on. But instead of subtracting 6 from 7, think of subtracting 60 from 79. The number of times 3 goes into 7 is 2, so the number of times it goes into 79 is 20. We keep the 20 in mind as part of the answer. Now our problem is 19 ÷ 3, which gives us 6 and a remainder of 1. The answer, then, is 26 with a remainder of 1.

We can do the problem 1234 ÷ 5 with the process used above or an easier method. Keep in mind that if we double both numbers in a division problem, the answer will stay the same. Thus, the problem 1234 ÷ 5 is the same as 2468 ÷ 10, and dividing by 10 is easy. The answer is 246.8.

With 2-digit division, our rapid 2-by-1 multiplication skills pay off. Let's determine the gas mileage if your car travels 353 miles on 14 gallons of gas. The problem is 353 ÷ 14; 14 goes into 35 twice, and 14 × 20 = 280. We keep the 20 in mind and subtract 280 from 353, which is 73. We now have a simpler division problem: 73 ÷ 14; the number of times 14 goes into 73 is 5 (14 × 5 = 70). The answer, then, is 25 with a remainder of 3.

Let's try 500 ÷ 73. How many times does 73 go into 500? It's natural to guess 7, but 7 × 73 = 511, which is a little too big. We now know that the quotient is 6, so we keep that in mind. We then multiply 6 × 73 to get 438, and using complements, we know that 500 – 438 = 62. The answer is 6 with a remainder of 62.

We can also do this problem another way. We originally found that 73 × 7 was too big, but we can take advantage of that calculation. We can think of the answer as 7 with a remainder of –11. That sounds a bit ridiculous, but it's the same as an answer of 6 with a remainder of 73 – 11 (= 62), and that agrees with our previous answer. This technique is called overshooting. With the problem 770 ÷ 79, we know that 79 × 10 = 790, which is too big by 20. Our first answer is 10 with a remainder of –20, but the final answer is 9 with a remainder of 79 – 20, which is 59.

A 4-digit number divided by a 2-digit number is about as large a mental division problem as most people can handle.

A 4-digit number divided by a 2-digit number is about as large a mental division problem as most people can handle. Consider the problem 2001 ÷ 23. We start with a 2-by-1 multiplication problem: 23 × 8 = 184; thus, 23 × 80 = 1840. We know that 80 will be part of the answer; now we subtract 2001 – 1840. Using complements, we find that 1840 is 160 away from 2000. Finally, we do 161 ÷ 23, and 23 × 7 = 161 exactly, which gives us 87 as the answer.

The problem 2012 ÷ 24 is easier. Both numbers here are divisible by 4; specifically, 2012 = 503 × 4, and 24 = 6 × 4. We simplify the problem to 503 ÷ 6, which reduces the 2-digit problem to a 1-digit division problem. The simplified problem gives us an answer of 83 5/6; as long as this answer and the one for 2012 ÷ 24 are expressed in fractions, they're the same.

To convert fractions to decimals, most of us know the decimal expansions when the denominator is 2, 3, 4, 5 or 10. The fractions with a denominator of 7 are the trickiest, but if you memorize the fraction for 1/7 (0.142857…), then you know the expansions for all the other sevenths fractions. The trick here is to think of drawing these numbers in a circle; you can then go around the circle to find the expansions for 2/7, 3/7, and so on. For example, 2/7 = 0.285714…, and 3/7 = 0.428571….

When dealing with fractions with larger denominators, we treat the fraction as a normal division problem, but we can occasionally take shortcuts, especially when the denominator is even. With odd denominators, you may not be able to find a shortcut unless the denominator is a multiple of 5, in which case you can double the numerator and denominator to make the problem easier.

Keep practicing the division techniques we've learned in this lecture, and you'll be dividing and conquering numbers mentally in no time. ■

Suggested Reading

Benjamin and Shermer, *Secrets of Mental Math: The Mathemagician's Guide to Lightning Calculation and Amazing Math Tricks*, chapter 5.

Julius, *More Rapid Math Tricks and Tips: 30 Days to Number Mastery.*

Kelly, *Short-Cut Math.*

Problems

Determine which numbers between 2 and 12 divide into each of the numbers below.

1. 4410

2. 7062

3. 2744

4. 33,957

Use the create-a-zero, kill-a-zero method to test the following.

5. Is 4913 divisible by 17?

6. Is 3141 divisible by 59?

7. Is 355,113 divisible by 7? Also do this problem using the special rule for 7s.

8. Algebraically, the divisibility rule for 7s says that $10a + b$ is a multiple of 7 if and only if the number $a - 2b$ is a multiple of 7. Explain why this works. (Hint: If $10a + b$ is a multiple of 7, then it remains a multiple of 7 after we multiply it by -2 and add $21a$. Conversely, if $a - 2b$ is a multiple of 7, then it remains so after we multiply it by 10 and add a multiple of 7.)

Mentally do the following 1-digit division problems.

9. $97 \div 8$

10. $63 \div 4$

11. $159 \div 7$

12. $4668 \div 6$

13. $8763 \div 5$

Convert the Fahrenheit temperatures below to Centigrade using the formula $C = (F - 32) \times 5/9$.

14. 80 degrees Fahrenheit

15. 65 degrees Fahrenheit

Mentally do the following 2-digit division problems.

16. $975 \div 13$

17. $259 \div 31$

18. 490 ÷ 62 (use overshooting)

19. 183 ÷ 19 (use overshooting)

Do the following division problems by first simplifying the problem to an easier division problem.

20. 4200 ÷ 8

21. 654 ÷ 36

22. 369 ÷ 45

23. 812 ÷ 12.5

24. Give the decimal expansions for 1/7, 2/7, 3/7, 4/7, 5/7, and 6/7.

25. Give the decimal expansion for 5/16.

26. Give the decimal expansion for 12/35.

27. When he was growing up, Professor Benjamin's favorite number was 2520. What is so special about that number?

Solutions for this lecture begin on page 103.

The Art of Guesstimation

Lecture 5

Your body is like a walking yardstick, and it's worth knowing things like the width of your hand from pinkie to thumb, or the size of your footsteps, or parts of your hand that measure to almost exactly one or two inches or one or two centimeters.

Mental estimation techniques give us quick answers to everyday questions when we don't need to know the answer to the last penny or decimal point. We estimate the answers to addition and subtraction problems by rounding, which can be useful when estimating the grocery bill. As each item is rung up, round it up or down to the nearest 50 cents.

To estimate answers to multiplication or division problems, it's important to first determine the order of magnitude of the answer. The general rules are as follows:

- For a multiplication problem, if the first number has x digits and the second number has y digits, then their product will have $x + y$ digits or, perhaps, $x + y - 1$ digits. Example: A 5-digit number times a 3-digit number creates a 7- or 8-digit number.

- To find out if the answer to $a \times b$ will have the larger or smaller number of digits, multiply the first digit of each number. If that product is 10 or more, then the answer will be the larger number. If that product is between 5 and 9, then the answer could go either way. If the product is 4 or less, then the answer will be the smaller number.

- For a division problem, the length of the answer is the difference of the lengths of the numbers being divided or 1 more. (Example: With an 8-digit number divided by a 3-digit number, the answer will have $8 - 3 = 5$ or 6 digits before the decimal point.)

- To find out how many digits come before the decimal point in the answer to $a \div b$, if the first digit of a is the same as the first digit of b, then compare the second digits of each number. If the first digit of a is larger than the first digit of b, then the answer will be the longer choice. If the first digit of a is less than the first digit of b, then the answer will be the shorter choice.

In estimating sales tax, if the tax is a whole number, such as 4%, then estimating it is just a straight multiplication problem. For instance, if you're purchasing a car for $23,456, then to estimate 4% tax, simply multiply 23,000 × 0.04 (= $920; exact answer: $938). If the tax is not a whole number, such as 4.5%, you can calculate it using 4%, but then divide that amount by 8 to get the additional 0.5%.

To estimate answers to multiplication or division problems, it's important to first determine the order of magnitude of the answer.

Suppose a bank offers an interest rate of 3% per year on its savings accounts. You can find out how long it will take to double your money using the "Rule of 70"; this calculation is 70 divided by the interest rate.

Suppose you borrow $200,000 to buy a house, and the bank charges an interest rate of 6% per year, compounded monthly. What that means is that the bank is charging you 6/12%, or 1/2%, interest for every month of your loan. If you have 30 years to repay your loan, how much will you need to pay each month? To estimate the answer, follow these steps:

- Find the total number of payments to be made: 30 × 12 = 360.

- Determine the monthly payment without interest: $200,000 ÷ 360. Simplify the problem by dividing everything by 10 (= 20,000 ÷ 36), then by dividing everything by 4 (= 5000 ÷ 9, or 1000 × 5/9). The fraction 5/9 is about 0.555, which means the monthly payment *without interest* would be about 1000 × 0.555, or $555.

- Determine the amount of interest owed in the first month: $\$200{,}000 \times 0.5\% = \$1000$.

A quick estimate of your monthly payment, then, would be $1000 to cover the interest plus $555 to go toward the principal, or $1555. This estimate will always be on the high side, because after each payment, you'll owe the bank slightly less than the original amount.

Square roots arise in many physical and statistical calculations, and we can estimate square roots using the divide-and-average method. To find the square root of a number, such as 40, start by taking any reasonable guess. We'll choose $6^2 = 36$. Next, divide 40 by 6, which is 6 with a remainder of 4, or 6 2/3. In other words, $6 \times 6\ 2/3 = 40$. The square root must lie between 6 and 6 2/3. If we average 6 and 6 2/3, we get 6 1/3, or about 6.33; the exact answer begins 6.32! ■

Important Term

square root: A number that, when multiplied by itself, produces a given number. For example, the square root of 9 is 3 and the square root of 2 begins 1.414.... Incidentally, the square root is defined to be greater than or equal to zero, so the square root of 9 is *not* –3, even though –3 multiplied by itself is also 9.

Suggested Reading

Benjamin and Shermer, *Secrets of Mental Math: The Mathemagician's Guide to Lightning Calculation and Amazing Math Tricks*, chapter 6.

Doerfler, *Dead Reckoning: Calculating Without Instruments.*

Hope, Reys, and Reys, *Mental Math in the Middle Grades.*

Kelly, *Short-Cut Math.*

Ryan, *Everyday Math for Everyday Life: A Handbook for When It Just Doesn't Add Up.*

Weinstein and Adam, *Guesstimation: Solving the World's Problems on the Back of a Cocktail Napkin.*

Problems

Estimate the following addition and subtraction problems by rounding each number to the nearest thousand, then to the nearest hundred.

1. 3764 + 4668

2. 9661 + 7075

3. 9613 – 1252

4. 5253 – 3741

Estimate the grocery total by rounding each number up or down to the nearest half dollar.

5.	6.	7.
5.24	0.87	0.78
0.42	2.65	1.86
2.79	0.20	0.68
3.15	1.51	2.73
0.28	0.95	4.29
0.92	2.59	3.47
4.39	1.60	2.65

What are the possible numbers of digits in the answers to the following problems?

8. 5 digits times 3 digits

9. 5 digits divided by 3 digits

10. 8 digits times 4 digits

11. 8 digits divided by 4 digits

For the following problems, determine the possible number of digits in the answers. (Some answers may allow two possibilities.) A number written as 3abc represents a 4-digit number with a leading digit of 3.

12. 3abc × 7def

13. 8abc × 1def

14. 2abc × 2def

15. 9abc ÷ 5de

16. 1abcdef ÷ 3ghij

17. 27abcdefg ÷ 26hijk

18. If a year has about 32 million seconds, then 1 trillion seconds is about how many years?

19. The government wants to buy a new weapons system costing $11 billion. The U.S. has about 100,000 public schools. If each school decides to hold a bake sale to raise money for the new weapons system, then about how much money does each school need to raise?

20. If an article is sent to two independent reviewers, and one reviewer finds 40 typos, the other finds 5 typos, and there were 2 typos in common, then estimate the total number of typos in the document.

21. Estimate 6% sales tax on a new car costing $31,500. Adjust your answer for 6.25% sales tax.

22. To calculate 8.5% tax, you can take 8% tax, then add the tax you just computed divided by what number? For 8.75% tax, you can take 9% tax, then subtract that tax divided by what number?

23. If money earns interest compounded at a rate of 2% per year, then about how many years would it take for that money to double?

24. Suppose you borrow $20,000 to buy a new car, the bank charges an annual interest rate of 3%, and you have 5 years to pay off the loan. Determine an underestimate and overestimate for your monthly payment, then determine the exact monthly payment.

25. Repeat the previous problem, but this time, the bank charges 6% annual interest and gives you 10 years to pay off the loan.

26. Use the divide-and-average method to estimate the square root of 27.

27. Use the divide-and-average method to estimate the square root of 153.

28. Speaking of 153, that's the first 3-digit number equal to the sum of the cubes of its digits ($153 = 1^3 + 5^3 + 3^3$). The next number with that property is 370. Can you find the third number with that property?

Solutions for this lecture begin on page 108.

Mental Math and Paper

Lecture 6

Even if you haven't been balancing your checkbook, you might now want to start. It's a great way to become more comfortable with numbers, and you'll understand exactly what's happening with your money!

In this lecture, we'll learn some techniques to speed up calculations done on paper, along with some interesting ways to check our answers. When doing problems on paper, it's usually best to perform the calculations from right to left, as we were taught in school. It's also helpful to say the running total as you go. To check your addition, add the numbers again, from bottom to top.

When doing subtraction on paper, we can make use of complements. Imagine balancing your checkbook; you start with a balance of $1235.79, from which you need to subtract $271.82. First, subtract the cents: 79 – 82. If the problem were 82 – 79, the answer would be 3 cents, but since it's 79 – 82, we take the complement of 3 cents, which is 97 cents. Next, we need to subtract 272, which we do by subtracting 300 (1235 – 300 = 935), then adding back its complement (28): 935 + 28 = 963. The new balance, then, is $963.97. We can check our work by turning the original subtraction problem into an addition problem.

Cross-multiplication is a fun way to multiply numbers of any length. This method is really just the distributive law at work. For example, the problem 23 × 58 is (20 + 3) × (50 + 8), which has four separate components: 20 × 50, 20 × 8, 3 × 50, and 3 × 8. The 3 × 8 we do at the beginning. The 20 × 50 we do at the end, and the 20 × 8 and 3 × 50 we do in **criss-cross** steps. If we extend this logic, we can do 3-by-3 multiplication or even higher. This method was first described in the book *Liber Abaci*, written in 1202 by Leonardo of Pisa, also known as Fibonacci.

The digit-sum check can be used to check the answer to a multiplication problem. Let's try the problem 314 × 159 = 49,926. We first sum the digits of the answer: $4 + 9 + 9 + 2 + 6 = 30$. We reduce 30 to a 2-digit number by adding its digits: $3 + 0 = 3$. Thus, the answer reduces to the number 3. Now, we reduce the original numbers: $314 \rightarrow 3 + 1 + 4 = 8$ and $159 \rightarrow 1 + 5 + 9 = 15$, which reduces to $1 + 5 = 6$. Multiply the reduced numbers, $8 \times 6 = 48$, then reduce that number: $4 + 8 = 12$, which reduces to $1 + 2 = 3$. The reduced numbers for both the answer and the problem match. If all the calculations are correct, then these numbers must match. Note that a match does not mean that your answer is correct, but if the numbers don't match, then you've definitely made a calculation error.

This method is also known as **casting out nines**, because when you reduce a number by summing its digits, the number you end up with is its remainder when divided by 9. For example, if we add the digits of 67, we get 13, and the digits of 13 add up to 4. If we take $67 \div 9$, we get 7 with a remainder of 4. Casting out nines also works for addition and subtraction problems, even those with decimals, and it may be useful for eliminating answers on standardized tests that do not allow calculators.

Casting out nines also works for addition and subtraction problems, even those with decimals, and it may be useful for eliminating answers on standardized tests that do not allow calculators.

The number 9, because of its simple multiplication table, its divisibility test, and the casting-out-nines process, seems almost magical. In fact, there's even a magical way to divide numbers by 9, using a process called Vedic division. This process is similar to the technique we learned for multiplying by 11 in Lecture 1, because dividing by 9 is the same as multiplying by 0.111111.

The **close-together method** can be used to multiply any two numbers that are near each other. Consider the problem 107×111. First, we note how far each number is from 100: 7 and 11. We then add either $107 + 11$ or $111 + 7$, both of which sum to 118. Next, we multiply 7×11, which is 77. Write the numbers down, and that's the answer: 11,877. The algebraic formula for this technique is $(z + a)(z + b) = (z + a + b)z + ab$, where typically, z is an easy number with zeros in it (such as $z = 100$ or $z = 10$) and a and b are the distances from the easy number. This technique also works for numbers below 100, but here, we use negative numbers for the distances from 100. Once you know how to do the close-together method on paper, it's not difficult to do it mentally; we'll try that in the next lecture. ■

Important Terms

casting out nines (also known as the method of digit sums): A method of verifying an addition, subtraction, or multiplication problem by reducing each number in the problem to a 1-digit number obtained by adding the digits. For example, 67 sums to 13, which sums to 4, and 83 sums to 11, which sums to 2. When verifying that $67 + 83 = 150$, we see that 150 sums to 6, which is consistent with $4 + 2 = 6$. When verifying $67 \times 83 = 5561$, we see that 5561 sums to 17 which sums to 8, which is consistent with $4 \times 2 = 8$.

close-together method: A method for multiplying two numbers that are close together. When the close-together method is applied to 23×26, we calculate $(20 \times 29) + (3 \times 6) = 580 + 18 = 598$.

criss-cross method: A quick method for multiplying numbers on paper. The answer is written from right to left, and nothing else is written down.

Suggested Reading

Benjamin and Shermer, *Secrets of Mental Math: The Mathemagician's Guide to Lightning Calculation and Amazing Math Tricks*, chapter 6.

Cutler and McShane, *The Trachtenberg Speed System of Basic Mathematics.*

Flansburg and Hay, *Math Magic: The Human Calculator Shows How to Master Everyday Math Problems in Seconds.*

Handley, *Speed Mathematics: Secrets of Lightning Mental Calculation.*

Julius, *More Rapid Math Tricks and Tips: 30 Days to Number Mastery.*

———, *Rapid Math Tricks and Tips: 30 Days to Number Power.*

Kelly, *Short-Cut Math.*

Problems

Add the following columns of numbers. Check your answers by adding the numbers in reverse order and by casting out nines.

1.	2.	3.
594	366	2.20
12	686	4.62
511	469	1.73
199	2010	32.30
3982	62	3.02
291	500	0.39
1697	4196	5.90

Do the following subtraction problems by first mentally computing the cents, then the dollars. Complements will often come in handy. Check your answers with an addition problem and with casting out nines.

4. 1776.65 – 78.95

5. 5977.31 – 842.78

6. 761.45 – 80.35

Use the criss-cross method to do the following multiplication problems. Verify that your answers are consistent with casting out nines.

7. 29 × 82

8. 764 × 514

9. 5593 × 2906

10. What is the remainder (not the quotient) when you divide 1,234,567 by 9?

11. What is the remainder (not the quotient) when you divide 12,345,678 by 9?

12. After doing the multiplication problem 1234 × 567,890, you get an answer that looks like 700,7#6,260, but the fifth digit is smudged, and you can't read it. Use casting out nines to determine the value of the smudged number.

Use the Vedic method to do the following division problems.

13. 3210 ÷ 9

14. 20,529 ÷ 9

15. 28,306 ÷ 9

16. 942,857 ÷ 9

Use the close-together method for the following multiplication problems.

17. 108 × 105

18. 92 × 95

19. 108 × 95

20. 998 × 997

21. 304 × 311

Solutions for this lecture begin on page 112.

Intermediate Multiplication

Lecture 7

The reason I like the factoring method is that it's easier on your memory, much easier than the addition or the subtraction method, because once you compute a number, ... you immediately put it to use.

In this lecture, we'll extend our knowledge of 2-by-1 and 3-by-1 multiplication to learn five methods for 2-by-2 multiplication. First is the addition method, which can be applied to any multiplication problem, although it's best to use it when at least one of the numbers being multiplied ends in a small digit. With this method, we round that number down to the nearest easy number. For 41 × 17, we treat 41 as 40 + 1 and calculate (40 × 17) + (1 × 17) = 680 + 17 = 697.

A problem like 53 × 89 could be done by the addition method, but it's probably easier to use the **subtraction method**. With this method, we treat 89 as 90 – 1 and calculate (53 × 90) – (53 × 1) = 4770 – 53 = 4717. The subtraction method is especially handy when one of the numbers ends in a large digit, such as 7, 8, or 9. Here, we round up to the nearest easy number. For 97 × 22, we treat 97 as 100 – 3, then calculate (100 × 22) – (3 × 22) = 2200 – 66 = 2134.

A third strategy for 2-by-2 multiplication is the factoring method. Again, for the problem 97 × 22, instead of rounding 97 up or rounding 22 down, we factor 22 as 11 × 2. We now have 97 × 11 × 2, and we can use the 11s trick from Lecture 1. The result for 97 × 11 is 1067; we multiply that by 2 to get 2134.

When using the factoring method, you often have several choices for how to factor, and you may wonder in what order you should multiply the factors. If you're quick with 2-by-1 multiplications, you can practice the "**math of least resistance**"—look at the problem both ways and take the easier path. The factoring method can also be used with decimals, such as when converting temperatures from Celsius to Fahrenheit.

Another strategy for 2-digit multiplication is **squaring**. For a problem like 13^2, we can apply the close-together method. We replace one of the 13s with 10; then, since we've gone down 3, we need to go back up by adding 3 to the other 13 to get 16. The first part of the calculation is now 10×16. To that result, we add the square of the number that went up and down (3): $10 \times 16 = 160$ and $160 + 3^2 = 169$. Numbers that end in 5 are especially easy to square using this method, as are numbers near 100.

If you're quick with 2-by-1 multiplications, you can practice the "math of least resistance"—look at the problem both ways and take the easier path.

Finally, our fifth mental multiplication strategy is the close-together method, which we saw in the last lecture. For a problem like 26×23, we first find a round number that is close to both numbers in the problem; we'll use 20. Next, we note how far away each of the numbers is from 20: 26 is 6 away, and 23 is 3 away. Now, we multiply 20×29. We get the number 29 in several ways: It's either $26 + 3$ or $23 + 6$; it comes from adding the original numbers together ($26 + 23 = 49$), then splitting that sum into $20 + 29$. After we multiply 29×20 ($= 580$), we add the product of the distances ($6 \times 3 = 18$): $580 + 18 = 598$.

After you've practiced these sorts of problems, you'll look for other opportunities to use the method. For example, for a problem like 17×76, you can make those numbers close together by doubling the first number and cutting the second number in half, which would leave you with the close-together problem 34×38.

The best method to use for mentally multiplying 2-digit numbers depends on the numbers you're given. If both numbers are the same, use the squaring method. If they're close to each other, use the close-together method. If one of the numbers can be factored into small numbers, use the factoring method. If one of the numbers is near 100 or it ends in 7, 8, or 9, try the subtraction method. If one of the numbers ends in a small digit, such as 1, 2, 3, or 4, or when all else fails, use the addition method. ■

Important Terms

math of least resistance: Choosing the easiest mental calculating strategy among several possibilities. For example, to do the problem 43×28, it is easier to do $43 \times 7 \times 4 = 301 \times 4 = 1204$ than to do $43 \times 4 \times 7 = 172 \times 7$.

squaring: Multiplying a number by itself. For example, the square of 5 is 25.

subtraction method: A method for multiplying numbers by turning the original problem into a subtraction problem. For example, $9 \times 79 = (9 \times 80) - (9 \times 1) = 720 - 9 = 711$, or $19 \times 37 = (20 \times 37) - (1 \times 37) = 740 - 37 = 703$.

Suggested Reading

Benjamin and Shermer, *Secrets of Mental Math: The Mathemagician's Guide to Lightning Calculation and Amazing Math Tricks*, chapter 3.

Kelly, *Short-Cut Math.*

Problems

Calculate the following 2-digit squares. Remember to begin by going up or down to the nearest multiple of 10.

1. 14^2

2. 18^2

3. 22^2

4. 23^2

5. 24^2

6. 25^2

7. 29^2

8. 31^2

9. 35^2

10. 36^2

11. 41^2

12. 44^2

13. 45^2

14. 47^2

15. 56^2

16. 64^2

17. 71^2

18. 82^2

19. 86^2

20. 93^2

21. 99^2

Do the following 2-digit multiplication problems using the addition method.

22. 31×23

23. 61×13

24. 52×68

25. 94×26

26. 47×91

Do the following 2-digit multiplication problems using the subtraction method.

27. 39×12

28. 79×41

29. 98×54

30. 87×66

31. 38×73

Do the following 2-digit multiplication problems using the factoring method.

32. 75×56

33. 67×12

34. 83×14

35. 79×54

36. 45×56

37. 68×28

Do the following 2-digit multiplication problems using the close-together method.

38. 13 × 19

39. 86 × 84

40. 77 × 71

41. 81 × 86

42. 98 × 93

43. 67 × 73

Do the following 2-digit multiplication problems using more than one method.

44. 14 × 23

45. 35 × 97

46. 22 × 53

47. 49 × 88

48. 42 × 65

Solutions for this lecture begin on page 116.

The Speed of Vedic Division

Lecture 8

There is more to Vedic mathematics than division, although we've seen much of it in this course already.

In this lecture, we explore strategies for doing division problems on paper that come to us from **Vedic mathematics**. With this approach, as we generate an answer, the digits of the answer play a role in generating more digits of the answer.

We first look at the processes of long and short division. Short division works well for 1-digit division and when dividing by a number between 10 and 20, but for numbers of 21 or greater, Vedic division is usually better. Vedic division is sort of like the subtraction method for multiplication applied to division.

Let's start with the problem 13,571 ÷ 39. The Vedic approach makes use of the fact that it's much easier to divide by 40 than 39. Dividing by 40 is essentially as easy as dividing by 4. If we divide 13,571 by 4, we get the 4-digit answer 3392.75. (We consider this a 4-digit answer because it has 4 digits before the decimal point.) If we're dividing 13,571 by 40, we simply shift the decimal point to get a 3-digit answer: 339.275.

With Vedic division, we start off as follows: 4 goes into 13, 3 times with a remainder of 1. The 3 goes above the line and the 1 goes next to the 5 below the line. Now, here's the twist: Instead of dividing 4 into 15, we divide 4 into 15 + 3; the 3 comes from the quotient digit above the line. We now have 15 + 3 = 18, and 4 goes into 18, 4 times with a remainder of 2. The 4 goes above the line, and the 2 goes next to the 7 below the line. Again, instead of dividing 4 into 27, we divide 4 into 27 + 4 (the quotient digit above the line), which is 31. We continue this process to get an answer to the original problem of 347 with a remainder of 38.

To see why this method works, let's look at the problem 246,810 ÷ 79. Essentially, when we divide by 79, we're dividing by 80 – 1, but if the process subtracts off three multiples of 80, it needs to add back three to compensate, just as we saw in the subtraction method for multiplication. The idea behind the Vedic method is that it's easier to divide by 80 than 79. For this problem, 80 goes into 246, 3 times, so we subtract 240, but we were supposed to subtract 3 × 79, not 3 × 80, so we have to add back 3 before taking the next step. Once we do this, we're at the same place we would be using long division.

Vedic division is sort of like the subtraction method for multiplication applied to division.

Sometimes, the division step results in a divisor that's greater than 10. If that happens, we carry the 10s digit into the previous column and keep going. For 1475 ÷ 29, we go up 1 to 30, so 3 is our divisor; 3 goes into 14, 4 times with a remainder of 2. The 4 goes above the line and the 2 goes next to the 7. Next, we do 3 into 27 + 4, which is 31; 3 goes into 31, 10 times with a remainder of 1. We write the 10 above the line, as before, making sure that the 1 goes in the previous column. When we reach the remainder step, we have to make sure to add 15 + 10, rather than 15 + 0. The result here is 50 with a remainder of 25.

If the divisor ends in 8, 7, 6, or 5, the procedure is almost the same. For the problem 123,456 ÷ 78, we go up 2 to get to 80 and use 8 as our divisor. Then, as we go through the procedure, we double the previous quotient at each step. If the original divisor ends in 7, we would add 3 to reach a round number, so at each step, we add 3 times the previous quotient. If the divisor ends in 6, we add 4 times the previous quotient, and if it ends in 5, we add 5 times the previous quotient. If the divisor ends in 1, 2, 3, or 4, we go down to reach a round number and *subtract* the previous quotient multiplied by that digit. In other words, the multiplier for these divisors would be –1, –2, –3, or –4. This subtraction step sometimes yields negative numbers; if this happens, we reduce the previous quotient by 1 and increase the remainder by the 1-digit divisor.

To get comfortable with Vedic division, you'll need to practice, but you'll eventually find that it's usually faster than short or long division for most 2-digit division problems. ■

Important Term

Vedic mathematics: A collection of arithmetic and algebraic shortcut techniques, especially suitable for pencil and paper calculations, that were popularized by Bhāratī Krishna Tirthajī in the 20th century.

Suggested Reading

Tekriwal, *5 DVD Set on Vedic Maths.*

Tirthajī, *Vedic Mathematics.*

Williams and Gaskell, *The Cosmic Calculator: A Vedic Mathematics Course for Schools, Book 3.*

Problems

Do the following 1-digit division problems on paper using short division.

1. 123,456 ÷ 7

2. 8648 ÷ 3

3. 426,691 ÷ 8

4. 21,472 ÷ 4

5. 374,476,409 ÷ 6

Do the following 1-digit division problems on paper using short division *and* by the Vedic method.

6. 112,300 ÷ 9

7. 43,210 ÷ 9

8. 47,084 ÷ 9

9. 66,922 ÷ 9

10. 393,408 ÷ 9

To divide numbers between 11 and 19, short division is very quick, especially if you can rapidly multiply numbers between 11 and 19 by 1-digit numbers. Do the following problems on paper using short division.

11. 159,348 ÷ 11

12. 949,977 ÷ 12

13. 248,814 ÷ 13

14. 116,477 ÷ 14

15. 864,233 ÷ 15

16. 120,199 ÷ 16

17. 697,468 ÷ 17

18. 418,302 ÷ 18

19. 654,597 ÷ 19

Use the Vedic method on paper for these division problems where the last digit is 9. The last two problems will have carries.

20. 123,456 ÷ 69

21. 14,113 ÷ 59

22. $71,840 \div 49$

23. $738,704 \div 79$

24. $308,900 \div 89$

25. $56,391 \div 99$

26. $23,985 \div 29$

27. $889,892 \div 19$

Use the Vedic method for these division problems where the last digit is 8, 7, 6, or 5. Remember that for these problems, the *multiplier* is 2, 3, 4, and 5, respectively.

28. $611,725 \div 78$

29. $415,579 \div 38$

30. $650,874 \div 87$

31. $821,362 \div 47$

32. $740,340 \div 96$

33. $804,148 \div 26$

34. $380,152 \div 35$

35. $103,985 \div 85$

36. Do the previous two problems by first doubling both numbers, then using short division.

Use the Vedic method for these division problems where the last digit is 1, 2, 3, or 4. Remember that for these problems, the multiplier is –1, –2, –3, and –4, respectively.

37. 113,989 ÷ 21

38. 338,280 ÷ 51

39. 201,220 ÷ 92

40. 633,661 ÷ 42

41. 932,498 ÷ 83

42. 842,298 ÷ 63

43. 547,917 ÷ 74

44. 800,426 ÷ 34

Solutions for this lecture begin on page 119.

Memorizing Numbers

Lecture 9

I can tell you from experience that if you use a list a lot, like, say, the presidents or a particular credit card number, then eventually, the phonetic code fades away and the numbers are converted to long-term memory, or you remember the numbers using other contextual information.

In this lecture, we'll learn a fun and amazingly effective way to memorize numbers. This skill will help you perform large calculations and help you memorize important numbers, such as credit card numbers. For most of this lecture, we'll take advantage of a phonetic code known as the **Major system**, which has been in the English language for nearly 200 years.

Here is the Major system: 1 = *t* or *d* sound; 2 = *n* sound; 3 = *m* sound; 4 = *r* sound; 5 = *L* sound; 6 = *ch*, *sh*, or *j* sound; 7 = *k* or *g* sound; 8 = *f* or *v* sound; 9 = *p* or *b* sound; and 0 = *s* or *z* sound.

After you've studied and memorized this phonetic code, you'll have an invaluable tool for turning numbers into words. We do this by inserting vowel sounds anywhere we'd like among the consonants. For example, suppose you need to remember the number 491. Using the phonetic code, you can turn this number into RABBIT, REPEAT, ORBIT, or another word by simply inserting vowels among the consonants in the code: 4 = *r*, 9 = *p* or *b*, and 1 = *t* or *d*. (Notice that even though RABBIT is spelled with two Bs, the *b* sound is pronounced only once. The number for RABBIT is 491, not 4991.) There are no digits for the consonants *h*, *w*, or *y*, so those can also be used whenever you'd like. Even though a number might have several words that represent it, each word can be turned into only one number. RABBIT, for example, represents only the number 491. Of course, we can also use the code in reverse to identify which number is represented by a particular word; for instance, PARTY would be 941.

The phonetic code is also useful for memorizing dates. For example, to remember that Andrew Jackson was elected president of the United States in 1828, we could turn 1828 into TFNF. You might picture Jackson as a TOUGH guy with a KNIFE. Or to remember that the Gettysburg Address was written in 1863 (TFJM), you might think that Lincoln wrote it to get out of a TOUGH JAM. On the Internet, you can find numerous sites that have converted entire dictionaries into phonetic code.

If you have a long number, such as a 16-digit credit card number, then it pays to look inside the number for particularly long words because the fewer words you use, the easier the resulting phrase is to remember. For the first 24 digits of pi, 3.14159265358979323846264, we can construct this sentence: "My turtle Poncho will, my love, pick up my new mover Ginger."

The phonetic code can also be used with the **peg system** to memorize any numbered list of objects. The peg system converts each number on the list into a tangible, easily visualized word called the peg word. My peg words for the numbers 1 though 10 are: tea, knee, moo, ear, oil, shoe, key, foe, pie, and dice. Notice that each of these words uses the sound for its corresponding number in the phonetic code. To remember that George Washington was the first president, I might picture myself drinking tea with him. Other associations might be a little bit strange, but that makes them even easier to remember.

You can use the phonetic code to provide more security to your computer password by adding extra digits.

If your list has more than 10 objects, then you need more peg words, and using the phonetic code, every 2-digit number can be turned into at least one word: 11 = tot, 12 = tin, 13 = tomb, and so on. My peg word for 40 is rose, and the phrase "red rose" reminds me that the 40th president was Ronald Reagan. I've also applied the peg system to learn where various elements appear on the periodic table.

You can use the phonetic code to provide more security to your computer password by adding extra digits. For instance, you might have one password that you like to use, such as BUNNY RABBIT, but you want to make slight alterations for each of your accounts. To adapt the password for your Visa account, you might attach the digits 80,741 (= VISA CARD).

The phonetic code is especially handy for numbers that you need to memorize for tests or for newly acquired phone numbers, addresses, parking spots, hotel rooms, and other numbers that you need to know for just a short while. I find the phonetic code to be useful for remembering partial answers when doing large mental calculations. We'll see more calculation examples that use mnemonics near the end of the course. ■

Important Terms

Major system: A phonetic code that assigns consonant sounds to digits. For example 1 gets the *t* or *d* sound, 2 gets the *n* sound, and so on. By inserting vowel sounds, numbers can be turned into words, which make them easier to remember. It is named after Major Beniowski, a leading memory expert in London, although the code was developed by Gregor von Feinagle and perfected by Aimé Paris.

peg system: A way to remember lists of objects, especially when the items of the list are given a number, such as the list of presidents, elements, or constitutional amendments. Each number is turned into a word using a phonetic code, and that word is linked to the object to be remembered.

Suggested Reading

Benjamin and Shermer, *Secrets of Mental Math: The Mathemagician's Guide to Lightning Calculation and Amazing Math Tricks*, chapter 9.

Higbee, *Your Memory: How It Works and How to Improve It.*

Lorayne and Lucas, *The Memory Book.*

Problems

Use the Major system to convert the following words into numbers.

1. News

2. Flash

3. Phonetic

4. Code

5. Makes

6. Numbers

7. Much

8. More

9. Memorable

For each of the numbers below, find at least two words for each number.

10. 11

11. 23

12. 58

13. 13

14. 21

15. 34

16. 55

17. 89

Use the phonetic code to create a mnemonic to remember the years of the following events.

18. Gutenberg operates first printing press in 1450.

19. Pilgrims arrive at Plymouth Rock in 1620.

20. Captain James Cook arrives in Australia in 1770.

21. Russian Revolution takes place in 1917.

22. First man sets foot on the Moon on July 21, 1969.

Create a mnemonic to remember the phone numbers listed below.

23. The Great Courses (in the U.S.): 800-832-2412

24. White House switchboard: 202-456-1414

25. Create your own personal set of peg words for the numbers 1 through 20.

26. How could you memorize the fact that the eighth U.S. president was Martin Van Buren?

27. How could you memorize the fact that the Fourth Amendment to the U.S. Constitution prohibits unreasonable searches and seizures?

28. How could you memorize the fact that the Sixteenth Amendment to the U.S. Constitution allows the federal government to collect income taxes?

Solutions for this lecture begin on page 128.

Calendar Calculating

Lecture 10

Sometimes people ask me the days of the week of ancient history, like what day of the week was January 1 in the year 0? The answer is "none of the above," since prior to the 3rd century, most places did not have seven days of the week. Instead, the situation was like what the Beatles once described as "Eight Days a Week."

In this lecture, we'll learn how to figure out the day of the week of any date in history. Once you've mastered this skill, you'll be surprised how often you use it. Starting with the year 2000, every year gets a code number. The code for 2000 is 0. The codes for Monday through Saturday are 1 through 6; the code for Sunday is 7 or 0. There are also codes for every month of the year: 6 (Jan.), 2 (Feb.), 2 (March), 5 (April), 0 (May), 3 (June), 5 (July), 1 (Aug.), 4 (Sept.), 6 (Oct.), 2 (Nov.), 4 (Dec.). In a **leap year**, January is 5 and February is 1.

It's helpful to develop a set of mnemonic devices to establish a link in your mind between each month and its code. For example, January might be associated with the word WINTER, which has the same number of letters as its code; February is the second month, and its code is 2; and so on.

To determine the day of the week for any year, we use this formula: month code + date + year code. For the date January 1, 2000, we go through these steps: The year 2000 was a leap year, so the month code for January is 5; add 1 for the date and 0 for the year. Those numbers sum to 6, which means that January 1, 2000, was a Saturday. If the sum of the codes and date is 7 or greater, we subtract the largest possible multiple of 7 to reduce it.

For the year 2001, the year code changes from 0 to 1; for 2002, it's 2; for 2003, it's 3; for 2004, because that's a leap year, the code is 5; and for 2005, the code is 6. The year 2006 would have a code of 7, but because we subtract 7s in the process of figuring out dates, we can subtract 7 here and simplify this code to 0.

The formula for determining the code for any year from 2000 to 2099 is: years + leaps – multiples of 7. Let's try the year 2025. We first plug the last two digits in for years. To figure out the leaps, recall that 2000 has a year code of 0. After that, the calendar will shift once for each year and once more for each leap year. By 2025, the calendar will have shifted 25 times for each year, plus once more for each leap year, and there are six leap years from 2001 to 2025 (years ÷ 4, ignoring any remainder). Thus, we add 25 + 6 = 31, then subtract the largest possible multiple of 7: 31 – 28 = 3, which is the year code for 2025.

Determining the year code is the hardest part of the calculation, so it helps to do that first.

Determining the year code is the hardest part of the calculation, so it helps to do that first. There is also a shortcut that comes in handy when the year ends in a high number. Between 1901 and 2099, the calendar repeats every 28 years. Thus, if you have a year such as 1998, you can subtract any multiple of 28 to make that number smaller, and the calendar will be exactly the same.

The general rule for leap years is that they occur every 4 years, with the exception that years divisible by 100 are not leap years. An exception to this exception is that if the year is divisible by 400, then it is still a leap year.

The year 1900 has a code of 1, 1800 is 3, 1700 is 5, and 1600 is 0. To determine the code for a year in the 1900s, the formula is years + leaps + 1 – multiples of 7; for the 1800s, years + leaps + 3 – multiples of 7; for the 1700s, years + leaps + 5 – multiples of 7; and for the 1600s, years + leaps – multiples of 7.

The calculations we've done all use the **Gregorian calendar**, which was established by Pope Gregory XIII in 1582 but wasn't universally adopted until the 1920s. Before the Gregorian calendar, European countries used the Julian calendar, established by Julius Caesar in 46 B.C. Under the Julian calendar, leap years happened every four years with no exceptions, but this created problems because the Earth's orbit around the Sun is not exactly 365.25 days. For this reason, we can't give the days of the week for dates in ancient history. ■

Important Terms

Gregorian calendar: Established by Pope Gregory XIII in 1582, it replaced the Julian calendar to more accurately reflect the length of the Earth's average orbit around the Sun; it did so by allowing three fewer leap years for every 400 years. Under the Julian calendar, every 4 years was a leap year, even when the year was divisible by 100.

leap year: A year with 366 days. According to our Gregorian calendar, a year is usually a leap year if it is divisible by 4. However, if the year is divisible by 100 and not by 400, then it is not a leap year. For example, 1700, 1800, and 1900 are not leap years, but 2000 is a leap year. In the 21st century, 2004, 2008, …, 2096 are leap years, but 2100 is not a leap year.

Suggested Reading

Benjamin and Shermer, *Secrets of Mental Math: The Mathemagician's Guide to Lightning Calculation and Amazing Math Tricks*, chapter 9.

Duncan, *The Calendar: The 5000-Year Struggle to Align the Clock and the Heavens—and What Happened to the Missing Ten Days.*

Reingold and Dershowitz, *Calendrical Calculations: The Millennium Edition.*

Problems

Here are the year codes for the years 2000 to 2040. The pattern repeats every 28 years (through 2099). For year codes in the 20th century, simply add 1 to the corresponding year code in the 21st century.

2000	2001	2002	2003	2004	2005	2006	2007	2008	2009	2010
0	1	2	3	5	6	0	1	3	4	5
	2011	2012	2013	2014	2015	2016	2017	2018	2019	2020
	6	1	2	3	4	6	0	1	2	4
	2021	2022	2023	2024	2025	2026	2027	2028	2029	2030
	5	6	0	2	3	4	5	0	1	2
	2031	2032	2033	2034	2035	2036	2037	2038	2039	2040
	3	5	6	0	1	3	4	5	6	1

1. Write down the month codes for each month in a leap year. How does the code change when it is not a leap year?

2. Explain why each year must always have at least one Friday the 13th and can never have more than three Friday the 13ths.

Determine the days of the week for the following dates. Feel free to use the year codes from the chart.

3. August 3, 2000

4. November 29, 2000

5. February 29, 2000

6. December 21, 2012

7. September 13, 2013

8. January 6, 2018

Calculate the year codes for the following years using the formula: year + leaps – multiple of 7.

9. 2020

10. 2033

11. 2047

12. 2074

13. 2099

Determine the days of the week for the following dates.

14. May 2, 2002

15. February 3, 2058

16. August 8, 2088

17. June 31, 2016

18. December 31, 2099

19. Determine the date of Mother's Day (second Sunday in May) for 2016.

20. Determine the date of Thanksgiving (fourth Thursday in November) for 2020.

For years in the 1900s, we use the formula: year + leaps + 1 – multiple of 7. Determine the year codes for the following years.

21. 1902

22. 1919

23. 1936

24. 1948

25. 1984

26. 1999

27. Explain why the calendar repeats itself every 28 years when the years are between 1901 and 2099. (Hint: Because 2000 is a leap year and a leap year occurs every 4 years, in a 28-year period, there will be exactly seven leap years.)

28. Use the 28-year rule to simplify the calculation of the year codes for 1984 and 1999.

Determine the days of the week for the following dates.

29. November 11, 1911

30. March 22, 1930

31. January 16, 1964

32. August 4, 1984

33. December 31, 1999

For years in the 1800s, the formula for the year code is years + leaps + 3 – multiple of 7. For years in the 1700s, the formula for the year code is years + leaps + 5 – multiple of 7. And for years in the 1600s, the formula for the year code is years + leaps – multiple of 7. Use this knowledge to determine the days of the week for the following dates from the Gregorian calendar.

34. February 12, 1809 (Birthday of Abe Lincoln *and* Charles Darwin)

35. March 14, 1879 (Birthday of Albert Einstein)

36. July 4, 1776 (Signing of the Declaration of Independence)

37. April 15, 1707 (Birthday of Leonhard Euler)

38. April 23, 1616 (Death of Miguel Cervantes)

39. Explain why the Gregorian calendar repeats itself every 400 years. (Hint: How many leap years will occur in a 400-year period?)

40. Determine the day of the week of January 1, 2100.

41. William Shakespeare and Miguel Cervantes both died on April 23, 1616, yet their deaths were 10 days apart. How can that be?

Solutions for this lecture begin on page 131.

Advanced Multiplication

Lecture 11

As I promised, these problems are definitely a challenge! As you saw, doing enormous problems ... requires all of the previous squaring and memory skills that we've learned. Once you can do a 4-digit square, even if it takes you a few minutes, the 3-digit squares suddenly don't seem so bad!

In this lecture, we'll look at mental math techniques for enormous problems, such as squaring 3- and 4-digit numbers and finding approximate cubes of 2-digit numbers. If you've been practicing the mental multiplication and squaring methods we've covered so far, you should be ready for this lecture.

To square 3-digit numbers quickly, you must be comfortable squaring 2-digit numbers. Let's start with 108^2. As we've seen before, we go down 8 to 100, up 8 to 116, then multiply $100 \times 116 = 11{,}600$; we then add 8^2 (= 64) to get 11,664. A problem like 126^2 becomes tricky if you don't know the 2-digit squares well, because you'll forget the first result ($152 \times 100 = 15{,}200$) while you try to work out 26^2. In this case, it might be helpful to say the 15,000, then raise 2 fingers (to represent 200) while you square 26.

Here's a geometry question: The Great Pyramid of Egypt has a square base, with side lengths of about 230 meters, or 755 feet. What is the area of the base? To find the answer in meters, we go down 30 to 200, up 30 to 260, then multiply $200 \times 260 = 52{,}000$; we then add 30^2 (= 900) to get 52,900 square meters.

To calculate the square footage (755^2), we could go up 45 to 800, then down 45 to 710, or we could use the push-together, pull-apart method: 755 + 755 = 1510, which can be pulled apart into 800 and 710. We now multiply $800 \times 710 = 568{,}000$, then add 45^2 (= 2025) to get 570,025 square feet.

One way to get better at 3-digit squares is to try 4-digit squares. In most cases, you'll need to use mnemonics for these problems. Let's try 2345^2. We go down 345 to 2000, up 345 to 2690. We then multiply 2000 × 2690, which is (2000 × 2600) + (2000 × 90) = 5,380,000. The answer will begin with 5,000,000, but the 380,000 is going to change.

How can we be sure that the 5,000,000 won't change? When we square a 4-digit number, the largest 3-digit number we will ever have to square in the middle is 500, because we always go up or down to the nearest thousand. The result of 500^2 is 250,000, which means that if we're holding onto a number that is less than 750,000 (here, 380,000), then we can be sure there won't be a carry.

One way to get better at 3-digit squares is to try 4-digit squares. In most cases, you'll need to use mnemonics for these problems.

How can we hold onto 380,000 while we square 345? Using the phonetic code we learned in Lecture 9, we send 380 to the MOVIES. Now we square 345: down 45 to 300, up 45 to 390; 300 × 390 = 117,000; add 45^2 (= 2025); and the result is 119,025. We hold onto the 025 by turning it into a SNAIL. We add 119,000 + MOVIES (380,000) = 499,000, which we can say. Then say SNAIL (025) for the rest of our answer. We've now said the answer: 5,499,025.

Notice that once you can square a 4-digit number, you can raise a 2-digit number to the 4th power just by squaring it twice. There's also a quick way to approximate 2-digit cubes. Let's try 43^3. We go down 3 to 40, down 3 to 40 again, then up 6 to 49. Our estimate of 43^3 is now 40 × 40 × 49. When we do the multiplication, we get an estimate of 78,400; the exact answer is 79,507.

Finally, we turn to 3-digit-by-2-digit multiplication. The easiest 3-by-2 problems have numbers that end in 0, because the 0s can be ignored until the end. Also easy are problems in which the 2-digit number can be factored into small numbers, which occurs about half the time. To find out how many hours are in a typical year, for example, we calculate 365 × 24, but 24 is 6 × 4, so we multiply 365 × 6, then multiply that result by 4.

The next easiest situation is when the 3-digit number can be factored into a 2-digit number × a 1-digit number. For instance, with 47 × 126, 47 is prime, but 126 is 63 × 2; we can multiply 47 × 63, then double that result. For the most difficult problems, we can break the 3-digit number into two parts and apply the distributive law. For a problem like 47 × 283, we multiply 47 × 280 and add 47 × 3.

In our last lecture, we'll see what you can achieve if you become seriously dedicated to calculation, and we'll consider broader benefits from what we've learned that are available to everyone. ■

Suggested Reading

Benjamin and Shermer, *Secrets of Mental Math: The Mathemagician's Guide to Lightning Calculation and Amazing Math Tricks*, chapter 8.

Doerfler, *Dead Reckoning: Calculating Without Instruments.*

Lane, *Mind Games: Amazing Mental Arithmetic Tricks Made Easy.*

Problems

Calculate the following 3-digit squares.

1. 107^2

2. 402^2

3. 213^2

4. 996^2

5. 396^2

6. 411^2

7. 155^2

8. 509^2

9. 320^2

10. 625^2

11. 235^2

12. 753^2

13. 181^2

14. 477^2

15. 682^2

16. 236^2

17. 431^2

Compute these 4-digit squares.

18. 3016^2

19. 1235^2

20. 1845^2

21. 2598^2

22. 4764^2

Raise these 2-digit numbers to the 4th power by squaring the number twice.

23. 20^4

24. 12^4

25. 32^4

26. 55^4

27. 71^4

28. 87^4

29. 98^4

Compute the following 3-digit-by-2-digit multiplication problems.

30. 864×20

31. 772×60

32. 140×23

33. 450×56

34. 860×84

35. 345×12

36. 456×18

37. 599×74

38. 753×56

39. 624×38

40. 349×97

41. 477×71

42. 181×86

43. 224×68

44. 241×13

45. 223×53

46. 682×82

Estimate the following 2-digit cubes.

47. 27^3

48. 51^3

49. 72^3

50. 99^3

51. 66^3

BONUS MATERIAL: We can also compute the exact value of a cube with only a little more effort. For example, to cube 42, we use $z = 40$ and $d = 2$. The approximate cube is $40 \times 40 \times 46 = 73{,}600$. To get the exact cube, we can use the following algebra: $(z + d)^3 = z(z(z + 3d) + 3d^2) + d^3$. First, we do $z(z + 3d) + 3d^2 = 40 \times 46 + 12 = 1852$. Then, we multiply this number by z again: $1852 \times 40 = 74{,}080$. Finally, we add $d^3 = 2^3 = 8$ to get 74,088.

Notice that when cubing a 2-digit number, in our first addition step, the value of $3d^2$ can be one of only five numbers: 3, 12, 27, 48, or 75. Specifically, if the number ends in 1 (so $d = 1$) or ends in 9 (so $d = -1$), then $3d^2 = 3$. Similarly, if the last digit is 2 or 8, we add 12; if it's 3 or 7, we add 27; if it's 4 or 6, we add 48; if it's 5, we add 75. Then, in the last step, we will always add or subtract one of five numbers, based on d^3. Here's the pattern:

If last digit is…	1	2	3	4	5	6	7	8	9
Adjust by…	+1	+8	+27	+64	+125	–64	–27	–8	–1

For example, what is the cube of 96? Here, $z = 100$ and $d = -4$. The approximate cube would be $100 \times 100 \times 88 = 880{,}000$. For the exact cube, we first do $100 \times 88 + 48 = 8848$. Then we multiply by 100 and subtract 64: $8848 \times 100 - 64 = 884{,}800 - 64 = 884{,}736$.

Using these examples as a guide, compute the exact values of the following cubes.

52. 13^3

53. 19^3

54. 25^3

55. 59^3

56. 72^3

Solutions for this lecture begin on page 137.

Masters of Mental Math

Lecture 12

You'll notice that cube rooting of 2-digit cubes doesn't really require much in the way of calculation. It's more like observation—looking at the number and taking advantage of a beautiful pattern.

We started this course using little more than the multiplication table, and we've since learned how to add, subtract, multiply, and divide enormous numbers. In this lecture, we'll review some of the larger lessons we've learned.

One of these lessons is that it pays to look at the numbers in a problem to see if they can somehow help to make the job of finding a solution easier. Can one of the numbers be broken into small factors; are the numbers close together; or can one of the numbers be rounded to give a good approximation of the answer?

We've also learned that difficult addition problems can often be made into easy subtraction problems and vice versa. In fact, if you want to become a "mental mathlete," it's useful to try to do problems in more than one way. We can approach a problem like 21 × 29, for example, using the addition, subtraction, factoring, or close-together methods.

We know that if we multiply a 5-digit number by a 3-digit number, the answer will have 8 (5 + 3) digits or maybe 7. If we pick the first digit of each number at random, then we would assume, just from knowing the multiplication table, that there's a good chance the product of those digits will be greater than 10, which would give us an 8-digit answer. According to **Benford's law**, however, it's far more likely that the original 5-digit and 3-digit numbers will start with a small number, such as 1, 2, or 3, which means that there's about a 50-50 chance of getting an answer with 8 digits or an answer with 7 digits. For most collections of numbers in the real world, such as street addresses or numbers on a tax return, there are considerably more numbers that start with 1 than start with 9.

Also in this course, we've learned how to apply the phonetic code to numbers that we have to remember and to use a set of codes to determine the day of the week for any date in the year. If anything, this course should have taught you to look at numbers differently, even when they don't involve a math problem.

As we've said, it usually pays to try to find features of problems that you can exploit. As an example, let's look at how to find the **cube root** of a number when the answer is a 2-digit number. Let's try 54,872; to find its cube root, all we need to know are the cubes of the numbers from 1 through 10. Notice that the last digits of these cubes are all different, and the last digit either matches the original number or is the 10s complement of the original number.

If anything, this course should have taught you to look at numbers differently, even when they don't involve a math problem.

To find the cube root of 54,872, we look at how the cube begins and ends. The number 54 falls between 3^3 and 4^3. Thus, we know that 54,000 falls between 30^3 (= 27,000) and 40^3 (= 64,000); its cube root must be in the 30s. The last digit of the cube is 2, and there's only one number from 1 to 10 whose cube ends in 2, namely, 8^3 (= 512); thus, the cube root of 54,872 is 38. Note that this method works only with perfect cubes.

Finally, we've learned that mental calculation is a process of constant simplification. Even very large problems can be broken down into simple steps. The problem $47{,}893^2$, for example, can be broken down into $47{,}000^2 + 893^2 + 47{,}000 \times 893 \times 2$. As we go through this problem, we make use of the criss-cross method, squaring smaller numbers, complements, and phonetic code—essentially, this is the math of least resistance.

To get into the *Guinness Book of World Records* for mental calculation, it used to be that contestants had to quickly determine the 13th root of a 100-digit number. To break the record now, contestants have to find the 13th root of a 200-digit number. Every two years, mathletes can also enter the Mental Calculation World Cup, which tests computation skills similar to what we've discussed in this course. Most of you watching this course are probably not aiming for these world championships, but the material we've covered should be useful to you throughout your life.

All mathematics begins with arithmetic, but it certainly doesn't end there. I encourage you to explore the joy that more advanced mathematics can bring in light of the experiences you've had with mental math. Some people lose confidence in their math skills at an early age, but I hope this course has given you the belief that you can do it. It's never too late to start looking at numbers in a new way. ■

Important Terms

Benford's law: The phenomenon that most of the numbers we encounter begin with smaller digits rather than larger digits. Specifically, for many real-world problems (from home addresses, to tax returns, to distances to galaxies), the first digit is N with probability $\log(N+1) - \log(N)$, where $\log(N)$ is the base 10 logarithm of N satisfying $10^{\log(N)} = N$.

cube root: A number that, when cubed, produces a given number. For example, the cube root of 8 is 2 since $2 \times 2 \times 2 = 8$.

Suggested Reading

Benjamin and Shermer, *Secrets of Mental Math: The Mathemagician's Guide to Lightning Calculation and Amazing Math Tricks*, chapters 8 and 9.

Doerfler, *Dead Reckoning: Calculating Without Instruments.*

Julius, *Rapid Math Tricks and Tips: 30 Days to Number Power.*

Lane, *Mind Games: Amazing Mental Arithmetic Tricks Made Easy.*

Rusczyk, *Introduction to Algebra.*

Smith, *The Great Mental Calculators: The Psychology, Methods and Lives of Calculating Prodigies Past and Present.*

Problems

We begin this section with a sample of review problems. Most likely, these problems would have been extremely hard for you to do before this course began, but I hope that now they won't seem so bad.

1. If an item costs $36.78, how much change would you get from $100?

2. Do the mental subtraction problem: $1618 - 789$.

Do the following multiplication problems.

3. 13×18

4. 65×65

5. 997×996

6. Is the number 72,534 a multiple of 11?

7. What is the remainder when you divide 72,534 by a multiple of 9?

8. Determine 23/7 to six decimal places.

9. If you multiply a 5-digit number beginning with 5 by a 6-digit number beginning with 6, then how many digits will be in the answer?

10. Estimate the square root of 70.

Do the following problems on paper and just write down the answer.

11. 509×325

12. $21{,}401 \div 9$

13. 34,567 ÷ 89

14. Use the phonetic code to memorize the following chemical elements: Aluminum is the 13th element; copper is the 29th element; and lead is the 82nd element.

15. What day of the week was March 17, 2000?

16. Compute 212^2.

17. Why must the cube root of a 4-, 5-, or 6-digit number be a 2-digit number?

Find the cube roots of the following numbers.

18. 12,167

19. 357,911

20. 175,616

21. 205,379

The next few problems will allow us to find the cube root when the original number is the cube of a 3-digit number. We'll first build up some ideas to find the cube root of 17,173,512, which is the cube of a 3-digit number.

22. Why must the first digit of the answer be 2?

23. Why must the last digit of the answer be 8?

24. How can we quickly tell that 17,173,512 is a multiple of 9?

25. It follows that the 3-digit number must be a multiple of 3 (because if the 3-digit number was not a multiple of 3, then its cube could not be a multiple of 9). What middle digits would result in the number 2_8 being a multiple of 3? There are three possibilities.

26. Use estimation to choose which of the three possibilities is most reasonable.

Using the steps above, we can do cube roots of any 3-digit cubes. The first digit can be determined by looking at the millions digits (the numbers before the first comma); the last digit can be determined by looking at the last digit of the cube; the middle digit can be determined through digit sums and estimation. There will always be three or four possibilities for the middle digit; they can be determined using the following observations, which you should verify.

27. Verify that if the digit sum of a number is 3, 6, or 9, then its cube will have digit sum 9.

28. Verify that if the digit sum of a number is 1, 4, or 7, then its cube will have digit sum 1.

29. Verify that if the digit sum of a number is 2, 5, or 8, then its cube will have digit sum 8.

Using these ideas, determine the 3-digit number that produces the cubes below.

30. Find the cube root of 212,776,173.

31. Find the cube root of 374,805,361.

32. Find the cube root of 4,410,944.

Compute the following 5-digit squares in your head!

33. $11{,}235^2$

34. $56{,}753^2$

35. $82{,}682^2$

Solutions for this lecture begin on page 142.

Solutions

Lecture 1

For later lectures, most of the solutions show how to generate the answer, but for Lecture 1, just the answers are shown below. Remember that it is just as important to hear the problem as to see the problem.

The following mental addition and multiplication problems can be done almost immediately, just by listening to the numbers from left to right.

1. 23 + 5 = 28

2. 23 + 50 = 73

3. 500 + 23 = 523

4. 5000 + 23 = 5023

5. 67 + 8 = 75

6. 67 + 80 = 147

7. 67 + 800 = 867

8. 67 + 8000 = 8067

9. $30 + 6 = 36$

10. $300 + 24 = 324$

11. $2000 + 25 = 2025$

12. $40 + 9 = 49$

13. $700 + 84 = 784$

14. $140 + 4 = 144$

15. $2500 + 20 = 2520$

16. $2300 + 58 = 2358$

17. $13 \times 10 = 130$

18. $13 \times 100 = 1300$

19. $13 \times 1000 = 13{,}000$

20. $243 \times 10 = 2430$

21. $243 \times 100 = 24{,}300$

22. $243 \times 1000 = 243{,}000$

23. 243×1 million $= 243$ million

24. Fill out the standard 10-by-10 multiplication table as quickly as you can. It's probably easiest to fill it out one row at a time by counting.

×	**1**	**2**	**3**	**4**	**5**	**6**	**7**	**8**	**9**	**10**
1	1	2	3	4	5	6	7	8	9	10
2	2	4	6	8	10	12	14	16	18	20
3	3	6	9	12	15	18	21	24	27	30
4	4	8	12	16	20	24	28	32	36	40
5	5	10	15	20	25	30	35	40	45	50
6	6	12	18	24	30	36	42	48	54	60
7	7	14	21	28	35	42	49	56	63	70
8	8	16	24	32	40	48	56	64	72	80
9	9	18	27	36	45	54	63	72	81	90
10	10	20	30	40	50	60	70	80	90	100

25. Create an 8-by-9 multiplication table in which the rows represent the numbers from 2 to 9 and the columns represent the numbers from 11 to 19. For an extra challenge, fill out the squares in random order.

×	**11**	**12**	**13**	**14**	**15**	**16**	**17**	**18**	**19**
2	22	24	26	28	30	32	34	36	38
3	33	36	39	42	45	48	51	54	57
4	44	48	52	56	60	64	68	72	76
5	55	60	65	70	75	80	85	90	95
6	66	72	78	84	90	96	102	108	114
7	77	84	91	98	105	112	119	126	133
8	88	96	104	112	120	128	136	144	152
9	99	108	117	126	135	144	153	162	171

26. Create the multiplication table in which the rows and columns represent the numbers from 11 to 19. For an extra challenge, fill out the rows in random order. Be sure to use the shortcuts we learned in this lecture, including those for multiplying by 11.

×	11	12	13	14	15	**16**	**17**	**18**	**19**
11	121	132	143	154	165	176	187	198	209
12	132	144	156	168	180	192	204	216	228
13	143	156	169	182	195	208	221	234	247
14	154	168	182	196	210	224	238	252	266
15	165	180	195	210	225	240	255	270	285
16	176	192	208	224	240	256	272	288	304
17	187	204	221	238	255	272	289	306	323
18	198	216	234	252	270	288	306	324	342
19	209	228	247	266	285	304	323	342	361

The following multiplication problems can be done just by listening to the answer from left to right.

27. $41 \times 2 = 82$

28. $62 \times 3 = 186$

29. $72 \times 4 = 288$

30. $52 \times 8 = 416$

31. $207 \times 3 = 621$

32. $402 \times 9 = 3618$

33. $543 \times 2 = 1086$

Do the following multiplication problems using the shortcut from this lecture.

34. $21 \times 11 = 231$ (since $2 + 1 = 3$, insert 3 between 2 and 1)

35. $17 \times 11 = 187$

36. $54 \times 11 = 594$

37. $35 \times 11 = 385$

38. $66 \times 11 = 726$ (since $6 + 6 = 12$, insert 2 between 6 and 6, then carry the 1)

39. $79 \times 11 = 869$

40. $37 \times 11 = 407$

41. $29 \times 11 = 319$

42. $48 \times 11 = 528$

43. $93 \times 11 = 1023$

44. $98 \times 11 = 1078$

45. $135 \times 11 = 1485$ (since $1 + 3 = 4$ and $3 + 5 = 8$)

46. $261 \times 11 = 2871$

47. $863 \times 11 = 9493$

48. $789 \times 11 = 8679$

49. Quickly write down the squares of all 2-digit numbers that end in 5.

$15^2 = 225$
$25^2 = 625$
$35^2 = 1225$
$45^2 = 2025$
$55^2 = 3025$
$65^2 = 4225$
$75^2 = 5625$
$85^2 = 7225$
$95^2 = 9025$

50. Since you can quickly multiply numbers between 10 and 20, write down the squares of the numbers 105, 115, 125, … 195, 205.

$105^2 = 11{,}025$
$115^2 = 13{,}225$
$125^2 = 15{,}625$
$135^2 = 18{,}225$
$145^2 = 21{,}025$
$155^2 = 24{,}025$
$165^2 = 27{,}225$
$175^2 = 30{,}625$
$185^2 = 34{,}225$
$195^2 = 38{,}025$
$205^2 = 42{,}025$

51. Square 995.

$995^2 = 990{,}025$.

52. Compute 1005^2.

1,010,025 (since $100 \times 101 = 10{,}100$; then attach 25)

Exploit the shortcut for multiplying 2-digit numbers that begin with the same digit and whose last digits sum to 10 to do the following problems.

53. $21 \times 29 = 609$ (using $2 \times 3 = 6$ and $1 \times 9 = 09$)

54. $22 \times 28 = 616$

55. $23 \times 27 = 621$

56. $24 \times 26 = 624$

57. $25 \times 25 = 625$

58. $61 \times 69 = 4209$

59. $62 \times 68 = 4216$

60. $63 \times 67 = 4221$

61. $64 \times 66 = 4224$

62. $65 \times 65 = 4225$

Lecture 2

Solve the following mental addition problems by calculating from left to right. For an *added* challenge, look away from the numbers after reading the problem.

1. $52 + 7 = 59$

2. $93 + 4 = 97$

3. $38 + 9 = 47$

4. $77 + 5 = 82$

5. $96 + 7 = 103$

6. $40 + 36 = 76$

7. $60 + 54 = 114$

8. $56 + 70 = 126$

9. $48 + 60 = 108$

10. $53 + 31 = 83 + 1 = 84$

11. $24 + 65 = 84 + 5 = 89$

12. $45 + 35 = 75 + 5 = 80$

13. $56 + 37 = 86 + 7 = 93$

14. $75 + 19 = 85 + 9 = 94$

15. $85 + 55 = 135 + 5 = 140$

16. $27 + 78 = 97 + 8 = 105$

17. $74 + 53 = 124 + 3 = 127$

18. $86 + 68 = 146 + 8 = 154$

19. $72 + 83 = 152 + 3 = 155$

Do these 2-digit addition problems in two ways; make sure the second way involves subtraction.

20. $68 + 97 = 158 + 7 = 165$
OR $68 + 97 = 68 + 100 - 3 = 168 - 3 = 165$

21. $74 + 69 = 134 + 9 = 143$
OR $74 + 69 = 74 + 70 - 1 = 144 - 1 = 143$

22. $28 + 59 = 78 + 9 = 87$
OR $28 + 59 = 28 + 60 - 1 = 88 - 1 = 87$

23. $48 + 93 = 138 + 3 = 141$
OR $48 + 93 = 48 + 100 - 7 = 148 - 7 = 141$
OR $48 + 93 = 93 + 50 - 2 = 143 - 2 = 141$

Try these 3-digit addition problems. The problems gradually become more difficult. For the harder problems, it may be helpful to say the problem out loud before starting the calculation.

24. $800 + 300 = 1100$

25. $675 + 200 = 875$

26. $235 + 800 = 1035$

27. $630 + 120 = 730 + 20 = 750$

28. $750 + 370 = 1050 + 70 = 1120$

29. $470 + 510 = 970 + 10 = 980$

30. $980 + 240 = 1180 + 40 = 1220$

31. $330 + 890 = 1130 + 90 = 1220$

32. $246 + 810 = 1046 + 10 = 1056$

33. 960 + 326 = 1260 + 26 = 1286

34. 130 + 579 = 679 + 30 = 709

35. 325 + 625 = 925 + 25 = 950

36. 575 + 675 = 1175 + 75 = 1100 + 150 = 1250

37. 123 + 456 = 523 + 56 = 573 + 6 = 579

38. 205 + 108 = 305 + 8 = 313

39. 745 + 134 = 845 + 34 = 875 + 4 = 879

40. 341 + 191 = 441 + 91 = 531 + 1 = 532
OR 341 + 200 – 9 = 541 – 9 = 532

41. 560 + 803 = 1360 + 3 = 1363

42. 566 + 185 = 666 + 85 = 746 + 5 = 751

43. 764 + 637 = 1364 + 37 = 1394 + 7 = 1401

Do the next few problems in two ways; make sure the second way uses subtraction.

44. 787 + 899 = 1587 + 99 = 1677 + 9 = 1686
OR 787 + 899 = 787 + 900 – 1 = 1687 – 1 = 1686

45. 339 + 989 = 1239 + 89 = 1319 + 9 = 1328
OR 339 + 989 = 339 + 1000 – 11 = 1339 – 11 = 1328

46. 797 + 166 = 897 + 66 = 957 + 6 = 963
OR 797 + 166 = 166 + 800 – 3 = 966 – 3 = 963

47. 474 + 970 = 1374 + 70 = 1444
OR 474 + 970 = 474 + 1000 – 30 = 1474 – 30 = 1444

Do the following subtraction problems from left to right.

48. $97 - 6 = 91$

49. $38 - 7 = 31$

50. $81 - 6 = 75$

51. $54 - 7 = 47$

52. $92 - 30 = 62$

53. $76 - 15 = 66 - 5 = 61$

54. $89 - 55 = 39 - 5 = 34$

55. $98 - 24 = 78 - 4 = 74$

Do these problems two different ways. For the second way, begin by subtracting too much.

56. $73 - 59 = 23 - 9 = 14$
OR $73 - 59 = 73 - (60 - 1) = 13 + 1 = 14$

57. $86 - 68 = 26 - 8 = 18$
OR $= 86 - (70 - 2) = 16 + 2 = 18$

58. $74 - 57 = 24 - 7 = 17$
OR $74 - 57 = 74 - (60 - 3) = 14 + 3 = 17$

59. $62 - 44 = 22 - 4 = 18$
OR $62 - (50 - 6) = 12 + 6 = 18$

Try these 3-digit subtraction problems, working from left to right.

60. $716 - 505 = 216 - 5 = 211$

61. 987 – 654 = 387 – 54 = 337 – 4 = 333

62. 768 – 222 = 568 – 22 = 548 – 2 = 546

63. 645 – 231 = 445 – 31 = 415 – 1 = 414

64. 781 – 416 = 381 – 16 = 371 – 6 = 365
OR 781 – 416 = 381 – 16 = 381 – (20 – 4) = 361 + 4 = 365

Determine the complements of the following numbers, that is, their distance from 100.

65. 100 – 28 = 72

66. 100 – 51 = 49

67. 100 – 34 = 66

68. 100 – 87 = 13

69. 100 – 65 = 35

70. 100 – 70 = 30

71. 100 – 19 = 81

72. 100 – 93 = 07

Use complements to solve these problems.

73. 822 – 593 = 822 – (600 – 7) = 222 + 7 = 229

74. 614 – 372 = 614 – (400 – 28) = 214 + 28 = 242

75. 932 – 766 = 932 – (800 – 34) = 132 + 34 = 166

76. 743 – 385 = 743 – (400 – 15) = 343 + 15 = 358

77. 928 – 262 = 928 – (300 – 38) = 628 + 38 = 666

78. 532 – 182 = 532 – (200 – 18) = 332 + 18 = 350

79. 611 – 345 = 611 – (400 – 55) = 211 + 55 = 226

80. 724 – 476 = 724 – (500 – 24) = 224 + 24 = 248

Determine the complements of these 3-digit numbers.

81. 1000 – 772 = 228

82. 1000 – 695 = 305

83. 1000 – 849 = 151

84. 1000 – 710 = 290

85. 1000 – 128 = 872

86. 1000 – 974 = 026

87. 1000 – 551 = 449

Use complements to determine the correct amount of change.

88. \$10 – \$2.71 = \$7.29

89. \$10 – \$8.28 = \$1.72

90. \$10 – \$3.24 = \$6.76

91. \$100 – \$54.93 = \$45.07

92. \$100 – \$86.18 = \$13.82

93. \$20 – \$14.36 = \$5.64

94. \$20 – \$12.75 = \$7.25

95. \$50 – \$31.41 = \$18.59

The following addition and subtraction problems arise while doing mental multiplication problems and are worth practicing before beginning Lecture 3.

96. 350 + 35 = 385

97. 720 + 54 = 774

98. 240 + 32 = 272

99. 560 + 56 = 616

100. 4900 + 210 = 5110

101. 1200 + 420 = 1620

102. 1620 + 48 = 1668

103. 7200 + 540 = 7740

104. 3240 + 36 = 3276

105. 2800 + 350 = 3150

106. 2150 + 56 = 2206

107. 800 – 12 = 788

108. 3600 – 63 = 3537

109. 5600 – 28 = 5572

110. 6300 – 108 = 6200 – 8 = 6192

Lecture 3

Calculate the following 2-by-1 multiplication problems in your head using the addition method.

1. $40 \times 8 = 320$

2. $42 \times 8 = 320 + 16 = 336$

3. $20 \times 4 = 80$

4. $28 \times 4 = 80 + 32 = 112$

5. $56 \times 6 = 300 + 36 = 336$

6. $47 \times 5 = 200 + 35 = 235$

7. $45 \times 8 = 320 + 40 = 360$

8. $26 \times 4 = 80 + 24 = 104$

9. $68 \times 7 = 420 + 56 = 476$

10. $79 \times 9 = 630 + 81 = 711$

11. $54 \times 3 = 150 + 12 = 162$

12. $73 \times 2 = 140 + 6 = 146$

13. $75 \times 8 = 560 + 40 = 600$

14. $67 \times 6 = 360 + 42 = 402$

15. $83 \times 7 = 560 + 21 = 581$

16. $74 \times 6 = 420 + 24 = 444$

17. $66 \times 3 = 180 + 18 = 198$

18. $83 \times 9 = 720 + 27 = 747$

19. $29 \times 9 = 180 + 81 = 261$

20. $46 \times 7 = 280 + 42 = 322$

Calculate the following 2-by-1 multiplication problems in your head using the addition method and the subtraction method.

21. $89 \times 9 = 720 + 81 = 801$
OR $89 \times 9 = (90 - 1) \times 9 = 810 - 9 = 801$

22. $79 \times 7 = 490 + 63 = 553$
OR $79 \times 7 = (80 - 1) \times 7 = 560 - 7 = 553$

23. $98 \times 3 = 270 + 24 = 294$
OR $98 \times 3 = (100 - 2) \times 3 = 300 - 6 = 294$

24. $97 \times 6 = 540 + 42 = 582$
OR $(100 - 3) \times 6 = 600 - 18 = 582$

25. $48 \times 7 = 280 + 56 = 336$
OR $48 \times 7 = (50 - 2) \times 7 = 350 - 14 = 336$

The following problems arise while squaring 2-digit numbers or multiplying numbers that are close together. They are essentially 2-by-1 problems with a 0 attached.

26. 20×16: $2 \times 16 = 20 + 12 = 32$, so $20 \times 16 = 320$

27. 20×24: $2 \times 24 = 40 + 8 = 48$, so $20 \times 24 = 480$

28. 20×25: $2 \times 25 = 50$, so $20 \times 25 = 500$

29. 20×26: $2 \times 26 = 40 + 12 = 52$, so $20 \times 26 = 520$

30. 20×28: $2 \times 28 = 40 + 16 = 56$, so $20 \times 2 = 560$

31. 20×30: 600

32. 30×28: $3 \times 28 = 60 + 24 = 84$, so $30 \times 28 = 840$

33. 30×32: $3 \times 32 = 90 + 6 = 96$, so $30 \times 32 = 960$

34. 40×32: $4 \times 32 = 120 + 8 = 128$, so $40 \times 32 = 1280$

35. 30×42: $3 \times 42 = 120 + 6 = 126$, so $30 \times 42 = 1260$

36. 40×48: $4 \times 48 = 160 + 32 = 192$, so $40 \times 48 = 1920$

37. 50×44: $5 \times 44 = 200 + 20 = 220$, so $50 \times 44 = 2200$

38. 60×52: $6 \times 52 = 300 + 12 = 312$, so $60 \times 52 = 3120$

39. 60×68: $6 \times 60 = 360 + 48 = 408$, so $60 \times 68 = 4080$

40. 60×69: $6 \times 69 = 360 + 54 = 414$, so $60 \times 69 = 4140$

41. 70×72: $7 \times 72 = 490 + 14 = 504$, so $70 \times 72 = 5040$

42. 70×78: $7 \times 78 = 490 + 56 = 546$, so $70 \times 78 = 5460$

43. 80×84: $8 \times 84 = 640 + 32 = 672$, so $80 \times 84 = 6720$

44. 80×87: $8 \times 87 = 640 + 56 = 696$, so $80 \times 87 = 6960$

45. 90×82: $9 \times 82 = 720 + 18 = 738$, so $90 \times 82 = 7380$

46. 90×96: $9 \times 96 = 810 + 54 = 864$, so $90 \times 96 = 8640$

Here are some more problems that arise in the first step of a 2-by-2 multiplication problem.

47. 30×23: $3 \times 23 = 60 + 9 = 69$, so $30 \times 23 = 690$

48. 60×13: $60 \times 13 = 60 + 18 = 78$, so $60 \times 13 = 780$

49. 50×68: $5 \times 68 = 300 + 40 = 340$, so $50 \times 68 = 3400$

50. 90×26: $9 \times 26 = 180 + 54 = 234$, so $90 \times 26 = 2340$

51. 90×47: $9 \times 47 = 360 + 63 = 423$, so $90 \times 47 = 4230$

52. 40×12: $4 \times 12 = 40 + 8 = 48$, so $40 \times 12 = 480$

53. 80×41: $8 \times 41 = 320 + 8 = 328$, so $80 \times 41 = 3280$

54. 90×66: $9 \times 66 = 540 + 54 = 594$, so $90 \times 66 = 5940$

55. 40×73: $4 \times 73 = 280 + 12 = 292$, so $40 \times 73 = 2920$

Calculate the following 3-by-1 problems in your head.

56. $600 \times 7 = 4200$

57. $402 \times 2 = 800 + 4 = 804$

58. $360 \times 6 = 1800 + 360 = 2160$

59. $360 \times 7 = 2100 + 420 = 2520$

60. $390 \times 7 = 2100 + 630 = 2730$

61. $711 \times 6 = 4200 + 66 = 4266$

62. $581 \times 2 = 1000 + 160 + 2 = 1162$

63. $161 \times 2 = 200 + 120 + 2 = 320 + 2 = 322$

64. $616 \times 7 = 4200 + (70 + 42) = 4200 + 112 = 4312$

65. $679 \times 5 = 3000$ (say it) $+ (350 + 45) = 3395$

66. $747 \times 2 = 1400$ (say it) $+ (80 + 14) = 1494$

67. $539 \times 8 = 4000$ (say it) $+ (240 + 72) = 4312$

68. $143 \times 4 = 400 + 160 + 12 = 560 + 12 = 572$

69. $261 \times 8 = 1600 + 480 + 8 = 2080 + 8 = 2088$

70. $624 \times 6 = 3600 + 120 + 24 = 3720 + 24 = 3744$

71. $864 \times 2 = 1600 + 120 + 8 = 1720 + 8 = 1728$

72. $772 \times 6 = 4200 + 420 + 12 = 4620 + 12 = 4632$

73. $345 \times 6 = 1800 + 240 + 30 = 2040 + 30 = 2070$

74. $456 \times 6 = 2400 + 300 + 36 = 2700 + 36 = 2736$

75. $476 \times 4 = 1600 + 280 + 24 = 1880 + 24 = 1904$

76. $572 \times 9 = 4500 + 630 + 18 = 5130 + 18 = 5148$

77. $667 \times 3 = 1800 + 180 + 21 = 1980 + 21 = 2001$

When squaring 3-digit numbers, the first step is to essentially do a 3-by-1 multiplication problem like the ones below.

78. 404×400: $404 \times 4 = 1616$, so $404 \times 400 = 161{,}600$

79. 226×200: $226 \times 2 = 400 + 52 = 452$, so $226 \times 200 = 45{,}200$

80. 422×400: $422 \times 4 = 1600 + 88 = 1688$, so $422 \times 400 = 168{,}800$

81. 110×200: $11 \times 2 = 22$, so $110 \times 200 = 22{,}000$

82. 518×500: $518 \times 5 = 2500 + 90 = 2590$, so $518 \times 500 = 259{,}000$

83. 340×300: $34 \times 3 = 90 + 12 = 102$, so $340 \times 300 = 102{,}000$

84. 650×600: $65 \times 6 = 360 + 30 = 390$, so $650 \times 600 = 390{,}000$

85. 270×200: $27 \times 2 = 40 + 14 = 54$, so $270 \times 200 = 54{,}000$

86. 706×800: $706 \times 8 = 5600 + 48 = 5648$, so $706 \times 800 = 564{,}800$

87. 162×200: $162 \times 2 = 200 + 120 + 4 = 320 + 4 = 324$, so $162 \times 200 = 32{,}400$

88. 454×500: $454 \times 5 = 2000$ (say it) $+ 250 + 20 = 2000 + 270 = 2270$, so $454 \times 500 = 227{,}000$

89. 664×700: $664 \times 7 = 4200 + 420 + 28 = 4620 + 28 = 4648$, so $664 \times 700 = 464{,}800$

Use the factoring method to multiply these 2-digit numbers together by turning the original problem into a 2-by-1 problem, followed by a 2-by-1 or 3-by-1 problem.

90. $43 \times 14 = 43 \times 7 \times 2 = (280 + 21) \times 2 = 301 \times 2 = 602$
OR $43 \times 14 = 43 \times 2 \times 7 = 86 \times 7 = 560 + 42 = 602$

91. $64 \times 15 = 64 \times 5 \times 3 = (300 + 20) \times 3 = 320 \times 3 = 900 + 60 = 960$

92. $75 \times 16 = 75 \times 8 \times 2 = (560 + 40) \times 2 = 600 \times 2 = 1200$

93. $54 \times 24 = 54 \times 6 \times 4 = (300 + 24) \times 4 = 324 \times 4 = 1200$ (say it) $+ (24 \times 4)$
$24 \times 4 = 80 + 16 = 96$, so $54 \times 24 = 1296$

94. $89 \times 72 = 89 \times 9 \times 8 = (720 + 81) \times 8 = 801 \times 8 = 6408$

In poker, there are 2,598,960 ways to be dealt 5 cards (from 52 different cards, where order is not important). Calculate the following multiplication problems that arise through counting poker hands.

95. The number of hands that are straights (40 of which are straight flushes) is $10 \times 4^5 = 4 \times 4 \times 4 \times 4 \times 4 \times 10 = 16 \times 4^3 \times 10$ $= 64 \times 4^2 \times 10 = 256 \times 4 \times 10 = 1024 \times 10 = 10{,}240$

96. The number of hands that are flushes is $(4 \times 13 \times 12 \times 11 \times 10 \times 9)/120$ $= 13 \times 11 \times 4 \times 9 = 143$ (close together) $\times 4 \times 9 = (400 + 160 + 12) \times 9$ $= 572 \times 9 = (4500 + 630 + 18) = 5130 + 18 = 5148$

97. The number of hands that are four-of-a-kind is $13 \times 48 = 13 \times 8 \times 6$ $= (80 + 24) \times 6 = 104 \times 6 = 624$

98. The number of hands that are full houses is $13 \times 12 \times 4 \times 6$ $= 156$ (close together) $\times 4 \times 6 = (400 + 200 + 24) \times 6 = 624 \times 6$ $= 3600 + 120 + 24 = 3720 + 24 = 3744$

Lecture 4

Determine which numbers between 2 and 12 divide into each of the numbers below.

1. 4410 is divisible by 2, 3, 5, 6, 7, 9, and 10.

Why? Last digit gives us 2, 5, and 10; digit sum = 9 gives us 3 and 9; divisible by 2 and 3 gives us divisibility by 6. Passes 7 test: $4410 \rightarrow 441 \rightarrow 44 - 2 = 42$ It fails tests for 4 (and, therefore, 8 and 12) and 11.

2. 7062 is divisible by 2, 3, 6, and 11.

Why? Last digit gives us 2; digit sum = 15 gives us 3; 2 and 3 imply 6; alternating sum of digits 7 – 0 + 6 – 2 = 11 gives us 11. Fails other tests.

3. 2744 is divisible by 2, 4, 7, and 8.

Why? 744 is divisible by 8; passes 7 test: 2744 → 274 – 8 = 266 → 26 – 12 = 14. Fails other tests.

4. 33,957 is divisible by 3, 7, 9, and 11.

Why? Digit sum = 27 gives us 3 and 9; passes 7 test: 33,957 → 3395 – 14 = 3381 → 338 – 2 = 336 → 33 – 12 = 21. Passes 11 test: 3 – 3 + 9 – 5 + 7 = 11. Fails other tests.

Use the create-a-zero, kill-a-zero method to test the following.

5. Is 4913 divisible by 17?

Yes, because 4913 → 4913 + 17 = 4930 → 493 → 493 + 17 = 510 → 51 is a multiple of 17.

6. Is 3141 divisible by 59?

No, because 3141 + 59 = 3200 → 320 → 32 is not a multiple of 59.

7. Is 355,113 divisible by 7?

No, because 355,113 + 7 = 355,120 → 35,512 → 35,512 + 28 =355,140→35,514→35,514–14=35,500→3550→355→355+35 = 390 → 39 is not a multiple of 7. Also, it fails the special rule for 7s: 355,113 – 6 = 355,107 → 35,510 – 14 = 35,496 → 3549 – 12 = 3537 → 353 – 14 = 339 → 33 – 18 = 15 is not a multiple of 7.

8. Algebraically, the divisibility rule for 7s says that $10a + b$ is a multiple of 7 if and only if the number $a - 2b$ is a multiple of 7. Explain why this works.

Suppose $10a + b$ is a multiple of 7, then it remains a multiple of 7 after we multiply it by –2, so $-20a - 2b$ will still be a multiple of 7. And since $21a$ is always a multiple of 7 (because it's $7 \times 3a$), we can add this to get $-20a - 2b + 21a$, which is $a - 2b$. So $a - 2b$ is still a multiple of 7.

Conversely, if $a - 2b$ is a multiple of 7, then it remains so after we multiply it by 10, so $10a - 20b$ is still a multiple of 7. Adding $21b$ (a multiple of 7) to this tells us that $10a + b$ is also a multiple of 7.

Mentally do the following 1-digit division problems.

9. $97 \div 8$

$$\begin{array}{r} 12 \text{ R } 1 = 12\,1/8 \\ 8\overline{)97} \\ -\underline{80} \\ 17 \\ -\underline{16} \\ 1 \end{array}$$

10. $63 \div 4$

$$\begin{array}{r} 15 \text{ R } 3 = 15\,3/4 \\ 4\overline{)63} \\ -\underline{40} \\ 23 \\ -\underline{20} \\ 3 \end{array}$$

11. 159 ÷ 7

$$\begin{array}{r} 22 \text{ R} 5 = 22\ 5/7 \\ 7\overline{)159} \\ -\underline{140} \\ 19 \\ -\underline{14} \\ 5 \end{array}$$

12. 4668 ÷ 6

$$\begin{array}{r} 778 \\ 6\overline{)4668} \\ -\underline{4200} \\ 468 \\ -\underline{420} \\ 48 \\ -\underline{48} \end{array}$$

13. 8763 ÷ 5 = (double both numbers) = 17,526 ÷ 10 = 1752.6

Convert the Fahrenheit temperatures below to Centigrade using the formula C = (F – 32) × 5/9.

14. 80 degrees Fahrenheit: (80 – 32) × 5/9 = 48 × 5/9 = 240 ÷ 9 = 80 ÷ 3 = 26 2/3 degrees Centigrade

15. 65 degrees Fahrenheit: (65 – 32) × 5/9 = 33 × 5/9 = 11 × 5/3 = 55 ÷ 3 = 18 1/3 degrees Centigrade

Mentally do the following 2-digit division problems.

16. 975 ÷ 13

$$\begin{array}{r} 75 \\ 13\overline{)975} \\ -\underline{910} \\ 65 \\ -\underline{65} \end{array}$$

17. $259 \div 31$

$$\begin{array}{r} 8 \text{ R}11 = 8\,11/31 \\ 31\overline{)259} \\ -\underline{248} \\ 11 \end{array}$$

18. $490 \div 62$ (use overshooting): $62 \times 8 = 496$, so $490 \div 62 = 8$ R -6 $= 7$ R 56

19. $183 \div 19$ (use overshooting): $19 \times 10 = 190$, so $183 \div 19 = 10$ R -7 $= 9$ R 12

Do the following division problems by first simplifying the problem to an easier division problem.

20. $4200 \div 8 = 2100 \div 4 = 1050 \div 2 = 525$

21. $654 \div 36$ (dividing both by 6) $= 109 \div 6 = 18\ 1/6$

22. $369 \div 45$ (doubling) $= 738 \div 90$; $738 \div 9 = 82$, so the answer is 8.2

23. $812 \div 12.5$ (doubling) $= 1624 \div 25 = 3248 \div 50 = 6496 \div 100 = 64.96$

24. Give the decimal expansions for 1/7, 2/7, 3/7, 4/7, 5/7, and 6/7.
1/7 = 0.142857 (repeated)
2/7 = 0.285714 (repeated)
3/7 = 0.428571 (repeated)
4/7 = 0.571428 (repeated)
5/7 = 0.714285 (repeated)
6/7 = 0.857142 (repeated)

25. Give the decimal expansion for 5/16: $50 \div 16 = 25 \div 8 = 3\ 1/8$ $= 3.125$, so $5/16 = 0.3125$

26. Give the decimal expansion for 12/35: $12/35 = 24 \div 70$. Given that $24/7 = 3\ 3/7 = 3.428571\ldots$, $12/35 = 0.3428571\ldots$

27. When he was growing up, Professor Benjamin's favorite number was 2520. What is so special about that number? It is the smallest positive number divisible by all the numbers from 1 to 10.

Lecture 5

Estimate the following addition and subtraction problems by rounding each number to the nearest thousand, then to the nearest hundred.

1. $3764 + 4668 \approx 4000 + 5000 = 9000$
OR $3764 + 4668 \approx 3800 + 4700 = 8500$

2. $9661 + 7075 \approx 10{,}000 + 7000 = 17{,}000$
OR $9661 + 7075 \approx 9700 + 7100 = 16{,}800$

3. $9613 - 1252 \approx 10{,}000 - 1000 = 9000$
OR $9613 - 1252 \approx 9600 - 1300 = 8300$

4. $5253 - 3741 \approx 5000 - 4000 = 1000$
OR $5253 - 3741 \approx 5300 - 3700 = 1600$

Estimate the grocery total by rounding each number up or down to the nearest half dollar.

5.	**6.**	**7.**
$5.24 \approx 5$	$0.87 \approx 1$	$0.78 \approx 1$
$0.42 \approx 0.5$	$2.65 \approx 2.5$	$1.86 \approx 2$
$2.79 \approx 3$	$0.20 \approx 0$	$0.68 \approx 0.5$
$3.15 \approx 3$	$1.51 \approx 1.5$	$2.73 \approx 2.5$
$0.28 \approx 0.5$	$0.95 \approx 1$	$4.29 \approx 4.5$
$0.92 \approx 1$	$2.59 \approx 2.5$	$3.47 \approx 3.5$
$4.39 \approx 4.5$	$1.60 \approx 1.5$	$2.65 \approx 2.5$
17.5	10.0	16.5

What are the possible numbers of digits in the answers to the following?

8. 5 digits times 3 digits is 7 or 8 digits.

9. 5 digits divided by 3 digits is 2 or 3 digits.

10. 8 digits times 4 digits is 11 or 12 digits.

11. 8 digits divided by 4 digits is 4 or 5 digits.

For the following problems, determine the possible number of digits in the answers. (Some answers may allow two possibilities.) A number written like 3abc represents a 4-digit number with leading digit of 3.

12. 3abc × 7def has 8 digits.

13. 8abc × 1def can have 7 or 8 digits.

14. 2abc × 2def has 7 digits.

15. 9abc ÷ 5de has 2 digits.

16. 1abcdef ÷ 3ghij has 2 digits.

17. 27abcdefg ÷ 26hijk has 4 digits.

18. If a year has about 32 million seconds, then 1 trillion seconds is about how many years?

The number 1 trillion has 13 digits, starting with 1, and 32 million has 8 digits, starting with 3, so 1 trillion divided by 32 million has 5 digits; thus, the answer is approximately 30,000.

19. The government wants to buy a new weapons system costing $11 billion. The U.S. has about 100,000 public schools. If each school decides to hold a bake sale to raise money for the new weapons system, then about how much money does each school need to raise?

The number 11 billion has 11 digits, starting with 11, and 100,000 has 6 digits, starting with 10, so the answer has $11 - 6 + 1 = 6$ digits, starting with 1; thus, the answer is about $110,000 per school.

20. If an article is sent to two independent reviewers, and one reviewer finds 40 typos, the other finds 5 typos, and there were 2 typos in common, then estimate the total number of typos in the document.

By Pólya's estimate, the total number of typos in the document is approximately $40 \times 5 \div 2 = 100$.

21. Estimate 6% sales tax on a new car costing $31,500. Adjust your answer for 6.25% sales tax.

$315 \times 6 = 1890$, so the sales tax is about $1900. For an additional 0.25%, increase this amount by $1900 ÷ 24 (since 6/24 = 0.25%), which is about $80; thus, the sales tax with the higher rate is about $1980.

22. To calculate 8.5% tax, you can take 8% tax, then add the tax you just computed divided by what number?

Since 8/16 = 0.5, you divide by 16.

For 8.75% tax, you can take 9% tax, then subtract that tax divided by what number?

To reduce the number by 0.25%, we divide the tax by 36, since 9/36 = 0.25.

23. If money earns interest compounded at a rate of 2% per year, then about how many years would it take for that money to double?

By the Rule of 70, since 70/2 = 35, it will take about 35 years to double.

24. Suppose you borrow \$20,000 to buy a new car, the bank charges an annual interest rate of 3%, and you have 5 years to pay off the loan. Determine an underestimate and overestimate for your monthly payment, then determine the exact monthly payment.

The number of monthly payments is $5 \times 12 = 60$. If no interest were charged, the monthly payment would be $20{,}000/60 \approx \$333$. But since the monthly interest is $3\%/12 = 0.25\%$, then you would owe $\$20{,}000(.25\%) = \$50$ in interest for the first month. The regular monthly payment would be, at most, $\$333 + \$50 = \$383$.

To get the exact monthly payment, we use the interest formula: $P \times i(1 + i)^m/((1 + i)^m - 1)$.

Here, $P = 20{,}000$, $i = 0.0025$, $m = 60$, and our calculator or search engine tells us $(1.0025)^{60} \approx 1.1616$; the monthly payment is about $\$20{,}000(.0025)(1.1616)/(0.1616) \approx \$359.40$/month, which is consistent with our lower bound of \$333 and our upper bound of \$383.

25. Repeat the previous problem, but this time, the bank charges 6% annual interest and gives you 10 years to pay off the loan.

The number of monthly payments is $10 \times 12 = 120$, so the lower estimate is $20{,}000/120 \approx \$167$/month. But since the monthly interest is $6\%/12 = 0.5\%$, then you would owe $\$20{,}000(.5\%) = \$100$ in interest for the first month. Thus, the regular monthly payment would be, at most, $\$167 + \$100 = \$267$. Plugging $P = 20{,}000$, $i = 0.005$, and $m = 120$ into the formula gives us $\$100(1.005)^{120}/((1.005)^{120} - 1) \approx \$181.94/(0.8194) \approx \$222$/month.

26. Use the divide-and-average method to estimate the square root of 27.

If we start with an estimate of 5, $27 \div 5 = 5.4$, and their average is 5.2. (Exact answer begins 5.196…)

27. Use the divide-and-average method to estimate the square root of 153.

If we start with an estimate of 12, 153 ÷ 12 = 12 9/12 = 12.75, and their average is 12.375. (Exact answer begins 12.369…)

28. Speaking of 153, that's the first 3-digit number equal to the sum of the cubes of its digits ($153 = 1^3 + 5^3 + 3^3$). The next number with that property is 370. Can you find the third number with that property?

Since $370 = 3^3 + 7^3 + 0^3$, it follows that $371 = 3^3 + 7^3 + 1^3$.

Lecture 6

Add the following columns of numbers. Check your answers by adding the numbers in reverse order and by casting out nines.

1.		2.		3.	
594	→ 9	366	→ 6	2.20	→ 4
12	→ 3	686	→ 2	4.62	→ 3
511	→ 7	469	→ 1	1.73	→ 2
199	→ 1	2010	→ 3	32.30	→ 8
3982	→ 4	62	→ 8	3.02	→ 5
291	→ 3	500	→ 5	0.39	→ 3
1697	→ 5	4196	→ 2	5.90	→ 5
7286	32	8289	27	50.16	30
|	|	|	|	|	|
5	5	9	9	3	3

Do the following subtraction problems by first mentally computing the cents, then the dollars. Complements will often come in handy. Check your answers with an addition problem and with casting out nines.

4. $1776.65 - 78.95 = 1697.70$ (Verifying, $1697.70 + 78.95 = 1776.65$)

$$5 \quad - \quad 2 \quad = \quad 3$$

5. $5977.31 - 842.78 = 5134.53$ (Verifying, $5134.53 + 842.78 = 5977.31$)

$$5 \quad - \quad 2 \quad = \quad 3$$

6. $761.45 - 80.35 = 681.10$ (Verifying, $681.10 + 80.35 = 761.45$)

$$5 \quad -7+9 = \quad 7$$

Use the criss-cross method to do the following multiplication problems. Verify that your answers are consistent with casting out nines.

7.

$$\begin{array}{r} 29 \rightarrow 11 \rightarrow \ 2 \\ \times \underline{82} \rightarrow 10 \rightarrow \times \underline{1} \\ 2378 \rightarrow 20 \rightarrow \ 2 \end{array}$$

8.

$$\begin{array}{r} 764 \rightarrow 17 \rightarrow \ 8 \\ \times \underline{514} \rightarrow 10 \rightarrow \times \underline{1} \\ 392{,}696 \rightarrow 35 \rightarrow \ 8 \end{array}$$

9.

$$\begin{array}{r} 5593 \rightarrow 22 \rightarrow \quad 4 \\ \times \underline{2906} \rightarrow 17 \rightarrow \times \underline{\ 8} \\ 16{,}253{,}258 \quad \rightarrow \quad 32 \end{array}$$

10. What is the remainder (not the quotient) when you divide 1,234,567 by 9?

Summing the digits, 1,234,567 → 28 → 10 → 1, so the remainder is 1.

11. What is the remainder (not the quotient) when you divide 12,345,678 by 9?

Summing the digits, 12,345,678 → 36 → 9, so the number is a multiple of 9, so dividing 12,345,678 by 9 yields a remainder of 0.

12. After doing the multiplication problem 1234 × 567,890, you get an answer that looks like 700,7#6,260, but the fifth digit is smudged, and you can't read it. Use casting out nines to determine the value of the smudged number.

Using digit sums, 1234 → 1 and 567,890 → 8, so their product must reduce to 1 × 8 = 8.

Summing the other digits, 7 + 0 + 0 + 7 + 6 + 2 + 6 + 0 = 28 → 1, so the smudged digit must be 7 in order to reach a total of 8.

Use the Vedic method to do the following division problems.

13. $3210 \div 9$

$$9\overline{)3\,2\,1\,0}\quad \text{quotient } 3\,5\,6 \;\; R\,6$$

14. $20{,}529 \div 9$

$$9\overline{)2\,0\,5\,2\,9}\quad \text{quotient } 2\,2\,7\,9 \;\; R\,18 = 2281\,R\,0$$

15. $28{,}306 \div 9$

$$9\overline{)2\,8\,3\,0\,6}\quad \text{quotient } \overset{1}{2}\,1\,4\,4 \;\; R\,10 = 3145\,R\,1$$

Solutions

16. $942{,}857 \div 9$

$$\begin{array}{l} \;\;1\;\;\;1\,1 \\ \;\;\underline{9\,4\,6\,5\,1}\;\; \text{R}\,8 = 104{,}761\,\text{R}\,8 \\ 9\,\overline{)9\,4\,2\,8\,5\,7} \end{array}$$

Use the close-together method for the following multiplication problems.

17.

$$\begin{array}{rr} 108 & (8) \\ \times\underline{105} & (5) \\ 113 & 40 \end{array}$$

18.

$$\begin{array}{rr} 92 & (-8) \\ \times\underline{95} & (-5) \\ 87 & 40 \end{array}$$

19.

$$\begin{array}{rrr} & 108 & (8) \\ & \times \quad \underline{95} & (-5) \\ 103 \times 100 = & 10{,}300 & \\ 8 \times (-5) = & \underline{-40} & \\ & 10{,}260 & \end{array}$$

20.

$$\begin{array}{rrr} & 998 & (-2) \\ & \times \quad \underline{997} & (-3) \\ 995 \times 1000 = & 995{,}000 & \\ (-2) \times (-3) = & \underline{+6} & \\ & 995{,}006 & \end{array}$$

21.

$$\begin{array}{rrr} & 304 & (4) \\ & \times \quad \underline{311} & (11) \\ 315 \times 300 = & 94{,}500 & \\ 4 \times 11 = & \underline{+44} & \\ & 94{,}544 & \end{array}$$

Lecture 7

Note: The details of many of the 2-by-1 and 3-by-1 multiplications are provided in the solutions for Lecture 3.

Calculate the following 2-digit squares. Remember to begin by going up or down to the nearest multiple of 10.

1. $14^2 = 10 \times 18 + 4^2 = 180 + 16 = 196$

2. $18^2 = 20 \times 16 + 2^2 = 320 + 4 = 324$

3. $22^2 = 20 \times 24 + 2^2 = 480 + 4 = 484$

4. $23^2 = 20 \times 26 + 3^2 = 520 + 9 = 529$

5. $24^2 = 20 \times 28 + 4^2 = 560 + 16 = 576$

6. $25^2 = 20 \times 30 + 5^2 = 600 + 25 = 625$

7. $29^2 = 30 \times 28 + 1^2 = 840 + 1 = 841$

8. $31^2 = 30 \times 32 + 1^2 = 960 + 1 = 961$

9. $35^2 = 30 \times 40 + 5^2 = 1200 + 25 = 1225$

10. $36^2 = 40 \times 32 + 4^2 = 1280 + 16 = 1296$

11. $41^2 = 40 \times 42 + 1^2 = 1680 + 1 = 1681$

12. $44^2 = 40 \times 48 + 4^2 = 1920 + 16 = 1936$

13. $45^2 = 40 \times 50 + 5^2 = 2000 + 25 = 2025$

14. $47^2 = 50 \times 44 + 3^2 = 2200 + 9 = 2209$

15. $56^2 = 60 \times 52 + 4^2 = 3120 + 16 = 3136$

16. $64^2 = 60 \times 68 + 4^2 = 4080 + 16 = 4096$

17. $71^2 = 70 \times 72 + 1^2 = 5040 + 1 = 5041$

18. $82^2 = 80 \times 84 + 2^2 = 6720 + 4 = 6724$

19. $86^2 = 90 \times 82 + 4^2 = 7380 + 16 = 7396$

20. $93^2 = 90 \times 96 + 3^2 = 8640 + 9 = 8649$

21. $99^2 = 100 \times 98 + 1^2 = 9800 + 1 = 9801$

Do the following 2-digit multiplication problems using the addition method.

22. $31 \times 23 = (30 + 1) \times 23 = (30 \times 23) + (1 \times 23) = 690 + 23 = 713$

23. $61 \times 13 = (60 + 1) \times 13 = (60 \times 13) + (1 \times 13) = 780 + 13 = 793$

24. $52 \times 68 = (50 + 2) \times 68 = (50 \times 68) + (2 \times 68) = 3400 + 136 = 3536$

25. $94 \times 26 = (90 + 4) \times 26 = (90 \times 26) + (4 \times 26) = 2340 + 104 = 2444$

26. $47 \times 91 = 47 \times (90 + 1) = (47 \times 90) + (47 \times 1) = 4230 + 47 = 4277$

Do the following 2-digit multiplication problems using the subtraction method.

27. $39 \times 12 = (40 - 1) \times 12 = 480 - 12 = 468$

28. $79 \times 41 = (80 - 1) \times 41 = 3280 - 41 = 3239$

29. $98 \times 54 = (100 - 2) \times 54 = 5400 - 108 = 5292$

30. $87 \times 66 = (90 - 3) \times 66 = (90 \times 66) - (3 \times 66) = 5940 - 198 = 5742$

31. $38 \times 73 = (40 - 2) \times 73 = (40 \times 73) - (2 \times 73) = 2920 - 146 = 2774$

Do the following 2-digit multiplication problems using the factoring method.

32. $75 \times 56 = 75 \times 8 \times 7 = 600 \times 7 = 4200$

33. $67 \times 12 = 67 \times 6 \times 2 = 402 \times 2 = 804$

34. $83 \times 14 = 83 \times 7 \times 2 = 581 \times 2 = 1162$

35. $79 \times 54 = 79 \times 9 \times 6 = 711 \times 6 = 4266$

36. $45 \times 56 = 45 \times 8 \times 7 = 360 \times 7 = 2520$

37. $68 \times 28 = 68 \times 7 \times 4 = 476 \times 4 = 1904$

Do the following 2-digit multiplication problems using the close-together method.

38. $13 \times 19 = (10 \times 22) + (3 \times 9) = 220 + 27 = 247$

39. $86 \times 84 = (80 \times 90) + (6 \times 4) = 7200 + 24 = 7224$

40. $77 \times 71 = (70 \times 78) + (7 \times 1) = 5460 + 7 = 5467$

41. $81 \times 86 = (80 \times 87) + (1 \times 6) = 6960 + 6 = 6966$

42. $98 \times 93 = (100 \times 91) + (-2 \times -7) = 9100 + 14 = 9114$

43. $67 \times 73 = (70 \times 70) + (-3 \times 3) = 4900 - 9 = 4891$

Do the following 2-digit multiplication problems using more than one method.

44. $14 \times 23 = 23 \times 7 \times 2 = 161 \times 2 = 322$
OR $14 \times 23 = 23 \times 2 \times 7 = 46 \times 7 = 322$
OR $14 \times 23 = (14 \times 20) + (14 \times 3) = 280 + 42 = 322$

45. $35 \times 97 = 35 \times (100 - 3) = 3500 - 35 \times 3 = 3500 - 105 = 3395$
OR $35 \times 97 = 97 \times 7 \times 5 = 679 \times 5 = 3395$

46. $22 \times 53 = 53 \times 11 \times 2 = 583 \times 2 = 1166$
OR $53 \times 22 = (50 + 3) \times 22 = 50 \times 22 + 3 \times 22 = 1100 + 66 = 1166$

47. $49 \times 88 = (50 - 1) \times 88 = (50 \times 88) - (1 \times 88) = 4400 - 88 = 4312$
OR $88 \times 49 = 88 \times 7 \times 7 = 616 \times 7 = 4312$
OR $49 \times 88 = 49 \times 11 \times 8 = 539 \times 8 = 4312$

48. $42 \times 65 = (40 \times 65) + (2 \times 65) = 2600 + 130 = 2730$
OR $65 \times 42 = 65 \times 6 \times 7 = 390 \times 7 = 2730$

Lecture 8

Do the following 1-digit division problems on paper using short division.

1. $123{,}456 \div 7$

$$\begin{array}{r|l} & 1\ 7\ 6\ 3\ 6\ \text{R}4 \\ \hline 7 & 1\,2_5 3_4 4_2 5_4 6 \end{array}$$

2. $8648 \div 3$

$$\begin{array}{r|l} & 2\ 8\ 8\ 2\ \text{R}2 \\ \hline 3 & 8_2 6_2 4_0 8 \end{array}$$

3. $426{,}691 \div 8$

$$\begin{array}{r|l} & 5\ 3\ 3\ 3\ 6\ \text{R}3 \\ \hline 8 & 4\,2_2 6_2 6_2 9_5 1 \end{array}$$

4. $21{,}472 \div 4$

$$\begin{array}{l} \ 5\,3\,6\,8\ \text{R}0 \\ 4\overline{)21_1 4_2 7_3 2} \end{array}$$

5. $374{,}476{,}409 \div 6$

$$\begin{array}{l} \ 6\,2\,4\,1\,2\,7\,3\,4\ \text{R}5 \\ 6\overline{)37_1 4_2 4_0 7_1 6_4 4_2 0_2 9} \end{array}$$

Do the following 1-digit division problems on paper using short division *and* by the Vedic method.

6. $112{,}300 \div 9$

$$\begin{array}{l} \ 1\,2\,4\,7\,7\ \text{R}7 \\ 9\overline{)11_2 2_4 3_7 0_7 0} \end{array} \qquad \text{Vedic:} \quad \begin{array}{l} 12477\ \ \text{R}7 \\ 9\overline{)112300} \end{array}$$

7. $43{,}210 \div 9$

$$\begin{array}{l} \ 4\,8\,0\,1\ \text{R}1 \\ 9\overline{)43_7 2_0 1_1 0} \end{array} \qquad \text{Vedic:} \quad \begin{array}{l} 1 \\ 4791\ \ \text{R}1 = 4801\ \text{R}1 \\ 9\overline{)43210} \end{array}$$

8. $47{,}084 \div 9$

$$\begin{array}{l} \ 5\,2\,3\,1\ \text{R}5 \\ 9\overline{)4\,7_2 0_2 8_1 4} \end{array} \qquad \text{Vedic:} \quad \begin{array}{l} 1\ \ 1 \\ 4221\ \ \text{R}5 = 5231\ \text{R}5 \\ 9\overline{)47084} \end{array}$$

9. $66{,}922 \div 9$

$$\begin{array}{l} \ 7\,4\,3\,5\ \ \text{R}7 \\ 9\overline{)66_3 9_3 2_5 2} \end{array} \qquad \text{Vedic:} \quad \begin{array}{l} 11 \\ 6335\ \ \text{R}7 = 7435\ \text{R}7 \\ 9\overline{)66922} \end{array}$$

10. $393{,}408 \div 9$

$$\begin{array}{l} \ 4\,3\,7\,1\,2\ \ \text{R}0 \\ 9\overline{)39_3 3_6 4_1 0_1 8} \end{array} \qquad \text{Vedic:} \quad \begin{array}{l} 1\ \ 1 \\ 33611\ \ \text{R}9 = 43{,}712\ \text{R}0 \\ 9\overline{)393408} \end{array}$$

To divide numbers between 11 and 19, short division is very quick, especially if you can rapidly multiply numbers between 11 and 19 by 1-digit numbers. Do the following problems on paper using short division.

11. 159,348 ÷ 11

$$\begin{array}{r} 1\ 4\ 4\ 8\ 6\ \text{R}2 \\ 11\overline{)1\,5_4 9_5 3_9 4_6 8} \end{array}$$

12. 949,977 ÷ 12

$$\begin{array}{r} 7\ 9\ 1\ 6\ 4\ \text{R}9 \\ 12\overline{)9\,4_{10} 9_1 9_7 7_5 7} \end{array}$$

13. 248,814 ÷ 13

$$\begin{array}{r} 1\ 9\ 1\ 3\ 9\ \text{R}7 \\ 13\overline{)2\,4_{11} 8_1 8_5 1_{12} 4} \end{array}$$

14. 116,477 ÷ 14

$$\begin{array}{r} 8\ 3\ 1\ 9\ \text{R}11 \\ 14\overline{)1\,1\,6_4 4_2 7_{13} 7} \end{array}$$

15. 864,233 ÷ 15

$$\begin{array}{r} 5\ 7\ 6\ 1\ 5\ \text{R}8 \\ 15\overline{)8\,6_{11} 4_9 2_2 3_8 3} \end{array}$$

16. 120,199 ÷ 16

$$\begin{array}{r} 7\ 5\ 1\ 2\ \text{R}7 \\ 16\overline{)1\,2\,0_8 1_1 9_3 9} \end{array}$$

17. 697,468 ÷ 17

$$\begin{array}{r} 4\ 1\ 0\ 2\ 7\ \text{R}9 \\ 17\overline{)6\,9_1 7_0 4_4 6_{12} 8} \end{array}$$

18. $418{,}302 \div 18$

$$\begin{array}{r|l} & \;2\;3\;2\;3\;\;9 \text{ R}0 \\ \hline 18 & 41_5 8_4 3_7 0_{16} 2 \end{array}$$

19. $654{,}597 \div 19$

$$\begin{array}{r|l} & \;3\;4\;4\;5\;2 \text{ R}9 \\ \hline 19 & 6\,5_8 4_8 5_9 9_4 7 \end{array}$$

Use the Vedic method on paper for these division problems where the last digit is 9. The last two problems will have carries.

20. $123{,}456 \div 69$

$$\begin{array}{r|l} & \;1\;7\;8\;9 \quad \text{R}15 \\ \hline 69 & 12_5 3_5 4_5 5_0 6 \\ +1 & \\ \hline 70 & \end{array}$$

First division step:	$12 \div 7 = 1 \text{ R } 5$
Second division step:	$(53 + 1) \div 7 = 7 \text{ R } 5$
Third division step:	$(54 + 7) \div 7 = 8 \text{ R } 5$
Fourth division step:	$(55 + 8) \div 7 = 9 \text{ R } 0$
Remainder:	$06 + 9 = 15$

21. $14{,}113 \div 59$

$$\begin{array}{r|l} & \;\;2\;3\;9 \quad \text{R}12 \\ \hline 59 & 1\,4_2 1_5 1_0 3 \\ +1 & \\ \hline 60 & \end{array}$$

22. $71{,}840 \div 49$

$$\begin{array}{r|l} & \;1\;4\;6\;6 \quad \text{R}6 \\ \hline 49 & 7_2 1_2 8_2 4_0 0 \\ +1 & \\ \hline 50 & \end{array}$$

23. 738,704 ÷ 79

$$
\begin{array}{l}
\;9\,3\,5\,0 \quad \text{R}\,54\\
79\overline{)73_1 8_3 7_0 0_5 4}\\
\underline{+1}\\
80
\end{array}
$$

24. 308,900 ÷ 89

$$
\begin{array}{l}
\;3\,4\,7\,0 \quad \text{R}\,70\\
89\overline{)30_3 8_5 9_0 0_7 0}\\
\underline{+1}\\
90
\end{array}
$$

25. 56,391 ÷ 99

$$
\begin{array}{l}
\;5\,6\,9 \quad \text{R}\,60\\
99\overline{)56_6 3_8 9_5 1}\\
\underline{+1}\\
100
\end{array}
$$

26. 23,985 ÷ 29

$$
\begin{array}{l}
\;1\\
\;7\,2\,6 \quad \text{R}\,31 = 826\ \text{R}\,31 = 827\ \text{R}\,2\\
29\overline{)23_2 9_0 8_2 5}\\
\underline{+1}\\
30
\end{array}
$$

First division step: 23 ÷ 3 = 7 R 2
Second division step: (29 + 7) ÷ 3 = 12 R 0
Third division step: (08 + 12) ÷ 3 = 6 R 2
Remainder: 25 + 6 = 31

27. 889,892 ÷ 19

$$
\begin{array}{l}
\;1\,1\\
\;4\,6\,7\,2\,5 \quad \text{R}\,27 = 46{,}835\ \text{R}\,27 = 46{,}836\ \text{R}\,8\\
19\overline{)8_0 8_0 9_1 8_1 9_1 2}\\
\underline{+1}\\
20
\end{array}
$$

First division step: $8 \div 2 = 4 \text{ R } 0$
Second division step: $(08 + 4) \div 2 = 6 \text{ R } 0$
Third division step: $(09 + 6) \div 2 = 7 \text{ R } 1$
Fourth division step: $(18 + 7) \div 2 = 12 \text{ R } 1$
Fifth division step: $(19 + 12) \div 2 = 15 \text{ R } 1$
Remainder: $12 + 15 = 27$

Use the Vedic method for these division problems where the last digit is 8, 7, 6, or 5. Remember that for these problems, the *multiplier* is 2, 3, 4, and 5, respectively.

28. 611,725 ÷ 78

$$\begin{array}{r|l} & 7\ 8\ 4\ 2 \quad \text{R } 49 \\ \hline 78 & 6\,1_5 1_1 7_1 2_4 5 \\ \underline{+2} & \\ 80 & \end{array}$$

First division step: $61 \div 8 = 7 \text{ R } 5$
Second division step: $(51 + 14) \div 8 = 8 \text{ R } 1$
Third division step: $(17 + 16) \div 8 = 4 \text{ R } 1$
Fourth division step: $(12 + 8) \div 8 = 2 \text{ R } 4$
Remainder: $45 + 4 = 49$

29. 415,579 ÷ 38

$$\begin{array}{r|l} & \quad\ 1\ 1 \\ & 1\ 0\ 8\ 2\ 5 \quad \text{R } 49 = 10{,}935 \text{ R } 49 = 10{,}936 \text{ R } 11 \\ \hline 38 & 4_0 1_3 5_3 5_3 7_1 9 \\ \underline{+2} & \\ 40 & \end{array}$$

First division step: $4 \div 4 = 1\ R\ 0$
Second division step: $(01 + 2) \div 4 = 0\ R\ 3$
Third division step: $(35 + 0) \div 4 = 8\ R\ 3$
Fourth division step: $(35 + 16) \div 4 = 12\ R\ 3$
Fifth division step: $(37 + 24) \div 4 = 15\ R\ 1$
Remainder = $19 + 30 = 49$

30. $650{,}874 \div 87$

$$\begin{array}{r|llllll l}
\multicolumn{1}{r}{} & & & & 1 & & & \\
\multicolumn{1}{r}{} & & 7 & 4 & 7 & 0 & & R114 = 7480\ R114 = 7481\ R\,27 \\ \cline{2-7}
87 & 6 & 5_2 & 0_5 & 8_7 & 7_8 & 4 & \\
\multicolumn{1}{r}{+3} \\ \cline{1-1}
\multicolumn{1}{r}{90}
\end{array}$$

31. $821{,}362 \div 47$

$$\begin{array}{r|lllllll l}
\multicolumn{1}{r}{} & & 1 & 7 & 4 & 7 & 5 & & R\,37 \\ \cline{2-8}
47 & 8_3 & 2_0 & 1_2 & 3_0 & 6_2 & 2 & & \\
\multicolumn{1}{r}{+3} \\ \cline{1-1}
\multicolumn{1}{r}{50}
\end{array}$$

32. $740{,}340 \div 96$

$$\begin{array}{r|llllll l}
\multicolumn{1}{r}{} & & & 1 & 1 & & & \\
\multicolumn{1}{r}{} & & 7 & 6 & 0 & 1 & & R\,84 = 7711\ R\,84 \\ \cline{2-7}
96 & 7 & 4_4 & 0_8 & 3_7 & 4_4 & 0 & \\
\multicolumn{1}{r}{+4} \\ \cline{1-1}
\multicolumn{1}{r}{100}
\end{array}$$

33. $804{,}148 \div 26$

$$\begin{array}{r|llllll l}
\multicolumn{1}{r}{} & & 1 & 2 & 4 & & & \\
\multicolumn{1}{r}{} & 2 & 9 & 6 & 8 & 2 & & R176 = 30{,}922\ R176 = 30{,}928\ R\,20 \\ \cline{2-7}
26 & 8_2 & 0_1 & 4_2 & 1_1 & 4_0 & 8 & \\
\multicolumn{1}{r}{+4} \\ \cline{1-1}
\multicolumn{1}{r}{30}
\end{array}$$

First division step: $8 \div 3 = 2 \text{ R } 2$
Second division step: $(20 + 8) \div 3 = 9 \text{ R } 1$
Third division step: $(14 + 36) \div 3 = 16 \text{ R } 2$
Fourth division step: $(21 + 64) \div 3 = 28 \text{ R } 1$
Fifth division step: $(14 + 112) \div 3 = 42 \text{ R } 0$
Remainder: $08 + 168 = 176$

Note: Problem 33 had many large carries, which can happen when the divisor is larger than the multiplier. Here, the divisor was small (3) and the multiplier was larger (4). Such problems might be better solved using short division.

34. 380,152 ÷ 35

$$\begin{array}{r|l} & \overset{1\ 2\ 3}{9\ 6\ 2\ 6} \quad \text{R}192 = 10{,}856\ \text{R}192 = 10{,}861\ \text{R}17 \\ 35 & 3\,8_2 0_1 1_3 5_1 2 \\ +5 & \\ \hline 40 & \end{array}$$

35. 103,985 ÷ 85

$$\begin{array}{r|l} & 1\ 2\ 2\ 3 \quad \text{R}30 \\ 85 & 1\,0_1 3_0 9_1 8_1 5 \\ +5 & \\ \hline 90 & \end{array}$$

36. Do the previous two problems by first doubling both numbers, then using short division.

380,152 ÷ 35 = 760,304 ÷ 70 =

$$\begin{array}{r|l} & 1\ 0\ 8\ 6\ 1.\ 4\ 6/7 = 10{,}861.48571428\ldots \\ 7 & 7_0 6_6 0_4 3_1 0._3 4 \end{array}$$

103,985 ÷ 85 = 207,970 ÷ 170 = 20,797 ÷ 17 =

$$\begin{array}{r|l} & 1\ 2\ 2\ 3\ \ 6/17 \\ 17 & 20_3 7_3 9_5 7 \end{array}$$

Solutions

Use the Vedic method for these division problems where the last digit is 1, 2, 3, or 4. Remember that for these problems, the multiplier is –1, –2, –3, and –4, respectively.

37. 113,989 ÷ 21

$$\begin{array}{r|l} & \ 5\ 4\ 2\ 8 \quad \text{R}1 \\ \hline 21 & 1\ 1_1 3_0 9_1 8_0 9 \\ \underline{-1} & \\ 20 & \end{array}$$

38. 338,280 ÷ 51

$$\begin{array}{r|l} & \ 6\ 6\ 3\ 3 \quad \text{R}-3 = 6632\ \text{R}\ 48 \\ \hline 51 & 3\ 3_3 8_2 2_1 8_0 0 \\ \underline{-1} & \\ 50 & \end{array}$$

39. 201,220 ÷ 92

$$\begin{array}{r|l} & \ 2\ 1\ 8\ 7 \quad \text{R}16 \\ \hline 92 & 2\ 0_2\ 1_8\ 2_8 2_3 0 \\ \underline{-2} & \\ 90 & \end{array}$$

40. 633,661 ÷ 42

$$\begin{array}{r|l} & 1\ 5\ 0\ 8\ 7 \quad \text{R}7 \\ \hline 42 & 6_2 3_1 3_3 6_4 6_2 1 \\ \underline{-2} & \\ 40 & \end{array}$$

Note: In the fourth division step, (36 – 0) ÷ 4 = 9 R 0 = 8 R 4.

41. 932,498 ÷ 83

$$\begin{array}{r|l} & 1\ 1\ 2\ 3\ 5 \quad \text{R}-7 = 11{,}234\ \text{R}\ 76 \\ \hline 83 & 9_1 3_2 2_3 4_4 9_0 8 \\ \underline{-3} & \\ 80 & \end{array}$$

42. 842,298 ÷ 63

$$\begin{array}{r|l} & \ 1\ 3\ 3\ 6\ 9 \quad R51 \\ \hline 63 & 8_2 4_3 2_5 2_7 9_7 8 \\ -3 & \\ \hline 60 & \end{array}$$

Note: In the fourth division step, (52 – 9) ÷ 6 = 7 R 1 = 6 R 7.

In the fifth division step, (79 – 18) ÷ 6 = 10 R 1 = 9 R 7.

43. 547,917 ÷ 74

$$\begin{array}{r|l} & \ \ 7\ 4\ 0\ 4 \quad R21 \\ \hline 74 & 5 4_5 7_1 9_3 1_3 7 \\ -4 & \\ \hline 70 & \end{array}$$

44. 800,426 ÷ 34

$$\begin{array}{r|l} & \ 2\ 3\ 5\ 4\ 1 \quad R32 \\ \hline 34 & 8_2 0_3 0_3 4_2 2_3 6 \\ -4 & \\ \hline 30 & \end{array}$$

Lecture 9

Use the Major system to convert the following words into numbers.

1. News = 20
2. Flash = 856
3. Phonetic = 8217
4. Code = 71
5. Makes = 370

6. Numbers = 23,940

7. Much = 36

8. More = 34

9. Memorable = 33,495

For each of the numbers below, find at least two words for each number. A few suggestions are given, but each number has more possibilities than those listed below.

10. 11 = date, diet, dot, dud, tot, tight, toot

11. 23 = name, Nemo, enemy, gnome, Nome

12. 58 = live, love, laugh, life, leaf, lava, olive

13. 13 = Adam, atom, dime, dome, doom, time, tome, tomb

14. 21 = nut, night, knight, note, ant, aunt, Andy, unit

15. 34 = mare, Homer, Mara, mere, meer, mire, and … more!

16. 55 = lily, Lola, Leila, Lyle, lolly, loyal, LOL

17. 89 = fib, fob, VIP, veep, Phoebe, phobia

Create a mnemonic to remember the years press in 1450.

He put it together using electric DRILLS!
He was TIRELESS in his efforts.

18. Pilgrims arrive at Plymouth Rock in 1620.

When they arrived, the pilgrims conducted a number of TEACH-INS.

A book about their voyage went through several EDITIONS.

19. Captain James Cook arrives in Australia in 1770.

The first animals he spotted were a DUCK and GOOSE.
For exercise, his crew would TAKE WALKS.

20. Russian Revolution takes place in 1917.

In the end, Lenin became TOP DOG, even though he was DIABETIC.

21. First man sets foot on the Moon on July 21, 1969.

The astronauts discovered CANDY (for 7/21) on TOP of their SHIP.
To get to sleep, the astronauts would COUNT DOPEY SHEEP.

Create a mnemonic to remember these phone numbers.

22. The Great Courses (in the U.S.): 800-832-2412

Their OFFICES experience FAMINE when a course is UNWRITTEN.
Their VOICES HAVE MANY a NEW ROUTINE.

23. White House switchboard: 202-456-1414

The president drives a NISSAN while eating RELISH and TARTAR.
The switchboard is run by an INSANE, REALLY SHY TRADER.

24. Create your own personal set of peg words for the numbers 1 through 20.

You'll have to do this one on your own!

25. How could you memorize the fact that the eighth U.S. president was Martin Van Buren?

Imagine a VAN BURNING that was caused by your FOE (named IVY or EVE?).

26. How could you memorize the fact that the Fourth Amendment to the U.S. Constitution prohibits unreasonable searches and seizures?

Imagine a soldier INSPECTING your EAR, which causes a SEIZURE. (Perhaps the solider was dressed like Julius Seizure, and he had gigantic EARs?)

27. How could you memorize the fact that the Sixteenth Amendment to the U.S. Constitution allows the federal government to collect income taxes?

This allowed the government to TOUCH all of our money!

Lecture 10

Here are the year codes for the years 2000 to 2040. The pattern repeats every 28 years (through 2099). For year codes in the 20th century, simply add 1 to the corresponding year code in the 21st century.

2000	2001	2002	2003	2004	2005	2006	2007	2008	2009	2010
0	1	2	3	5	6	0	1	3	4	5
	2011	2012	2013	2014	2015	2016	2017	2018	2019	2020
	6	1	2	3	4	6	0	1	2	4
	2021	2022	2023	2024	2025	2026	2027	2028	2029	2030
	5	6	0	2	3	4	5	0	1	2
	2031	2032	2033	2034	2035	2036	2037	2038	2039	2040
	3	5	6	0	1	3	4	5	6	1

1. Write down the month codes for each month in a leap year. How does the code change when it is not a leap year?

If it is not a leap year, the month codes are (from January to December) 622 503 514 624.
In a leap year, the code for January changes to 5 and February changes to 1.

2. Explain why each year must always have at least one Friday the 13th and can never have more than three Friday the 13ths.

This comes from the fact that in every year (whether or not it's a leap year), all seven month codes, 0 through 6, are used at least once, and no code is used more than three times. For example, if it is not a leap year and the year had three Friday the 13ths, they must have occurred in February, March, and November (all three months have the same month code of 2). In a leap year, this can only happen for the months of January, April, and July (with the same month code of 5).

Determine the days of the week for the following dates. Feel free to use the year codes from the chart.

3. August 3, 2000 = month code + date + year code – multiple of 7
= 1 + 3 + 0 = 4 = Thursday

4. November 29, 2000 = 2 + 29 + 0 – 28 = 3 = Wednesday

5. February 29, 2000 = 1 + 29 + 0 – 28 = 2 = Tuesday

6. December 21, 2012 = 4 + 21 + 1 – 21 = 5 = Friday

7. September 13, 2013 = 4 + 13 + 2 – 14 = 5 = Friday

8. January 6, 2018 = 6 + 6 + 1 = 13 – 7 = 6 = Saturday

Calculate the year codes for the following years using the formula: year + leaps – multiple of 7.

9. 2020: Since leaps = $20 \div 4 = 5$, the year code is $20 + 5 - 21 = 4$.

10. 2033: Since leaps = $33 \div 4 = 8$ (with remainder 1, which we ignore), the year code is $33 + 8 - 35 = 6$.

11. 2047: year code = $47 + 11 - 56 = 2$

12. 2074: year code = $74 + 18 - 91 = 1$ (or $74 + 18 - 70 - 21 = 1$)

13. 2099: year code = $99 + 24 - 119 = 4$ (or $99 + 24 - 70 - 49 = 4$)

Determine the days of the week for the following dates.

14. May 2, 2002: year code = 2;
month + date + year code – multiple of 7 = $0 + 2 + 2 = 4$ = Thursday

15. February 3, 2058: year code = $58 + 14 - 70 = 2$;
day = $2 + 3 + 2 - 7 = 0$ = Sunday

16. August 8, 2088: year code = $88 + 22 - 105 = 5$;
day = $1 + 8 + 5 - 14 = 0$ = Sunday

17. June 31, 2016: Ha! This date doesn't exist! But the calculation would produce an answer of $3 + 31 + 6 - 35 = 5$ = Friday.

18. December 31, 2099: year code = 4 (above);
day = $4 + 31 + 4 - 35 = 4$ = Thursday

19. Determine the date of Mother's Day (second Sunday in May) for 2016.

The year 2016 has year code 6, and May has month code 0. $6 + 0 = 6$. To reach Sunday, we must get a total of 7 or 14 or 21. … The first Sunday is May 1 (since $6 + 1 = 7$), so the second Sunday is May 8.

20. Determine the date of Thanksgiving (fourth Thursday in November) for 2020.

The year 2020 has year code 4, and November has month code 2: $4 + 2 = 6$. To reach Thursday, we must get a day code of 4 or 11 or 18. … Since $6 + 5 = 11$, the first Thursday in November will be November 5. Thus, the fourth Thursday in November is November $5 + 21 =$ November 26.

For years in the 1900s, we use the formula: year + leaps + 1 – multiple of 7. Determine the year codes for the following years.

21. 1902: year code $= 2 + 0 + 1 - 0 = 3$

22. 1919: year code $= 19 + 4 + 1 - 21 = 3$

23. 1936: year code $= 36 + 9 + 1 - 42 = 4$

24. 1948: year code $= 48 + 12 + 1 - 56 = 5$

25. 1984: year code $= 84 + 21 + 1 - 105 = 1$

26. 1999: year code $= 99 + 24 + 1 - 119 = 5$ (This makes sense because the following year, 2000, is a leap year, which has year code $5 + 2 - 7 = 0$.)

27. Explain why the calendar repeats itself every 28 years when the years are between 1901 and 2099.

Between 1901 and 2099, a leap year occurs every 4 years, even when it includes the year 2000. Thus any 28 consecutive between 1901 and 2099 will contain exactly 7 leap years. Hence, in a 28-year period, the calendar will shift 28 for each year plus 7 more times for each leap year for a total shifting of 35 days. Because 35 is a multiple of 7, the days of the week stay the same.

28. Use the 28-year rule to simplify the calculation of the year codes for 1984 and 1999.

For 1984, we subtract $28 \times 3 = 84$ from 1984. Thus, 1984 has the same year code as 1900, which has year code 1.

For 1999, we subtract 84 to get 1915, which has year code $15 + 3 + 1 - 14 = 5$.

Determine the days of the week for the following dates.

29. November 11, 1911: year code $= 11 + 2 + 1 - 14 = 0$;
day $= 2 + 11 + 0 - 7 = 6 =$ Saturday

30. March 22, 1930: year code $= 30 + 7 + 1 - 35 = 3$;
day $= 2 + 22 + 3 - 21 = 6 =$ Saturday

31. January 16, 1964: year code $= 64 + 16 + 1 - 77 = 4$;
day = 5 (leap year) $+ 16 + 4 - 21 = 4 =$ Thursday

32. August 4, 1984: year code = 1 (above);
day $= 1 + 4 + 1 = 6 =$ Saturday

33. December 31, 1999: year code = 5 (above);
day $= 4 + 31 + 5 - 35 = 5 =$ Friday

For years in the 1800s, the formula for the year code is years + leaps + 3 – multiple of 7. For years in the 1700s, the formula for the year code is years + leaps + 5 – multiple of 7. And for years in the 1600s, the formula for the year code is years + leaps – multiple of 7. Use this knowledge to determine the days of the week for the following dates from the Gregorian calendar.

34. February 12, 1809 (Birthday of Abe Lincoln *and* Charles Darwin):
year code $= 9 + 2 + 3 - 14 = 0$; day $= 2 + 12 + 0 - 14 = 0 =$ Sunday.

35. March 14, 1879 (Birthday of Albert Einstein):
year code = 79 + 19 + 3 – 98 = 3; day = 2 + 14 + 3 – 14 = 5 = Friday.

36. July 4, 1776 (Signing of the Declaration of Independence):
year code = 76 + 19 + 5 – 98 = 2; day = 5 + 4 + 2 – 7 = 4 = Thursday.

37. April 15, 1707 (Birthday of Leonhard Euler):
year code = 7 + 1 + 5 – 7 = 6; day = 5 + 15 + 6 – 21 = 5 = Friday.

38. April 23, 1616 (Death of Miguel Cervantes):
year code = 16 + 4 – 14 = 6; day = 5 + 23 + 6 – 28 = 6 = Saturday.

39. Explain why the calendar repeats itself every 400 years in the Gregorian calendar. (Hint: how many leap years will occur in a 400-year period?)

In a 400-year period, the number of leap years is 100 – 3 = 97. (Recall that in the next 400 years, 2100, 2200, and 2300 are not leap years, but 2400 is a leap year.) Hence, the calendar will shift 400 times (once for each year) plus 97 more times (for each leap year), for a total of 497 shifts. Because 497 is a multiple of 7 (= 7 × 71), the day of the week will be the same.

40. Determine the day of the week of January 1, 2100.

This day will be the same as January 1, 1700 (not a leap year), which has year code 5; hence, the day of the week will be 6 + 1 + 5 – 7 = 5 = Friday; this is consistent with our earlier calculation that December 31, 2099 is a Thursday.

41. William Shakespeare and Miguel Cervantes both died on April 23, 1616, yet their deaths were 10 days apart. How can that be?

Cervantes was from Spain, which adopted the Gregorian calendar. England, Shakespeare's home, was still on the Julian calendar, which was 10 days "behind" the Gregorian calendar. When Shakespeare died on the Julian date of April 23, 1616, the Gregorian date was May 3, 1616.

Lecture 11

Calculate the following 3-digit squares. Note that most of the 3-by-1 multiplications appear in the problems and solutions to Lecture 3, and most of the 2-digit squares appear in the problems and solutions to Lecture 7.

1. $107^2 = 100 \times 114 + 7^2 = 11{,}400 + 49 = 11{,}449$

2. $402^2 = 400 \times 404 + 2^2 = 161{,}600 + 4 = 161{,}604$

3. $213^2 = 200 \times 226 + 13^2 = 45{,}200 + 169 = 45{,}369$

4. $996^2 = 1000 \times 992 + 4^2 = 992{,}000 + 16 = 992{,}016$

5. $396^2 = 400 \times 392 + 4^2 = 156{,}800 + 16 = 156{,}816$

6. $411^2 = 400 \times 422 + 11^2 = 168{,}800 + 121 = 168{,}921$

7. $155^2 = 200 \times 110 + 45^2 = 22{,}000 + 2025 = 24{,}025$

8. $509^2 = 500 \times 518 + 9^2 = 259{,}000 + 81 = 259{,}081$

9. $320^2 = 300 \times 340 + 20^2 = 102{,}000 + 400 + 102{,}400$

10. $625^2 = 600 \times 650 + 25^2 = 390{,}000 + 625 = 390{,}625$

11. $235^2 = 200 \times 270 + 35^2 = 54{,}000 + 1{,}225 = 55{,}225$

12. $753^2 = 800 \times 706 + 47^2 = 564{,}800 + 2{,}209 = 567{,}009$

13. $181^2 = 200 \times 162 + 19^2 = 32{,}400 + 361 = 32{,}761$

14. $477^2 = 500 \times 454 + 23^2 = 227{,}000 + 529 = 227{,}529$

15. $682^2 = 700 \times 664 + 18^2 = 464{,}800 + 324 = 465{,}124$

16. $236^2 = 200 \times 272 + 36^2 = 54{,}400 + 1{,}296 = 55{,}696$

17. $431^2 = 400 \times 462 + 31^2 = 184{,}800 + 961 = 185{,}761$

Compute these 4-digit squares. Note that all of the required 3-digit squares have been solved in the exercises above. After the first multiplication, you can usually say the millions digit; the displayed word is the phonetic representation of the underlined number. Also, some of these calculations require 4-by-1 multiplications; these are indicated after the solution.

18. $3016^2 = 3000 \times 3032 + 16^2 = 9{,}096{,}000 + 256 = 9{,}096{,}256$

(Note: $3 \times 3032 = 3 \times 3000 + 3 \times 32 = 9000 + 96 = 9096$)

19. $1235^2 = 1000 \times 1470 + 235^2 = 1{,}\underline{470}{,}000$ (ROCKS) + $55{,}\underline{225}$ (NO NAIL) = 1,525,225

20. $1845^2 = 2000 \times 1690 + 155^2 = 3{,}\underline{380}{,}000$ (MOVIES) + $24{,}\underline{025}$ (SNAIL) = 3,404,025

(Note: $2 \times 169 = 200 + 120 + 18 = 320 + 18 = 338$, so $2 \times 1690 = 3{,}380$. Note also that the number 1690 can be found by doubling 1845, giving 3690, which splits into 2000 and 1690.)

21. $2598^2 = 3000 \times 2196 + 402^2 = 6,\underline{588},000$ (LOVE OFF) $+ 161,\underline{604}$ (CHASER) $= 6,749,604$

(Note: $3 \times 2196 = 3 \times 2000 + 3 \times 196 = 6000 + (300 + 270 + 18) = 6000 + (570 + 18) = 6588$. Note also that $2598 \times 2 = 5196 = 3000 + 2196$.)

22. $4764^2 = 5000 \times 4528 + 236^2 = 22,\underline{640},000$ (CHAIRS) $+ 55,\underline{696}$ (SHEEPISH) $= 22,695,696$

(Note: $5 \times 4528 = 5 \times 4500 + 5 \times 28 = 22,500 + 140 = 22,640$. Note also that $4764 \times 2 = 9528 = 5000 + 4528$.)

Raise these two-digit numbers to the 4th power by squaring the number twice.

23. $20^4 = 400^2 = 160,000$

24. $12^4 = 144^2 = 100 \times 188 + 44^2 = 18,800 + 1,936 = 20,736$

25. $32^4 = 1024^2 = 1000 \times 1048 + 24^2 = 1,048,000 + 576 = 1,048,576$

26. $55^4 = 3025^2 = 3000 \times 3050 + 25^2 = 9,150,000 + 625 = 9,150,625$

27. $71^4 = 5041^2 = 5000 \times 5082 + 41^2 = 25,\underline{410},000$ (ROADS) $+ 1,\underline{681}$ (SHIFT) $= 25,411,681$

28. $87^4 = 7569^2 = 8000 \times 7138 + 431^2 = 57,\underline{104},000$ (TEASER) $+ 185,\underline{761}$ (CASHED) $= 57, 289,761$

(Note: $8 \times 7138 = 8 \times 7100 + 8 \times 38 = 56,800 + 304 = 57,104$. Also note that the number 7138 can be obtained by doing $7569 \times 2 = 15,138$ so that the numbers being multiplied are 8000 and 7138.)

29. $98^4 = 9604^2 = 10,000 \times 9208 + 396^2 = 92,\underline{080},000$ (SAVES) $+ 156,\underline{816}$ (FOOTAGE) $= 92,236,816$

(Note: $9604 \times 2 = 19,208 = 10,000 + 9208$.)

Compute the following 3-digit-by-2-digit multiplication problems. Note that many of the 3-by-1 calculations appear in the solutions to Lecture 3, and many of the 2-by-2 calculations appear in the solutions to Lecture 7.

30. $864 \times 20 = 17{,}280$

31. $772 \times 60 = 46{,}320$

32. $140 \times 23 = 23 \times 7 \times 2 \times 10 = 161 \times 2 \times 10 = 322 \times 10 = 3220$

33. $450 \times 56 = 450 \times 8 \times 7 = 3600 \times 7 = 25{,}200$

34. $860 \times 84 = 86 \times 84 \times 10 = 7224 \times 10 = 72{,}240$

35. $345 \times 12 = 345 \times 6 \times 2 = 2070 \times 2 = 4140$

36. $456 \times 18 = 456 \times 6 \times 3 = 2736 \times 3 = 8100 + 108 = 8208$

37. $599 \times 74 = (600 - 1) \times 74 = 44{,}400 - 74 = 44{,}326$

38. $753 \times 56 = 753 \times 8 \times 7 = 6024 \times 7 = 42{,}000 + 168 = 42{,}168$

39. $624 \times 38 = 38 \times 104 \times 6 = (3800 + 152) \times 6 = 3952 \times 6 = 23{,}400 + 312 = 23{,}712$

40. $349 \times 97 = 349 \times (100 - 3) = 34{,}900 - 1047 = 33{,}853$

41. $477 \times 71 = (71 \times 400) + (71 \times 77) = 28{,}400 + 5467 = 33{,}867$

42. $181 \times 86 = (100 \times 86) + (81 \times 86) = 8600 + 6966 = 15{,}566$

43. $224 \times 68 = 68 \times 8 \times 7 \times 4 = 544 \times 7 \times 4 = 3808 \times 4 = 15{,}232$

44. $241 \times 13 = (13 \times 24 \times 10) + (13 \times 1) = 3120 + 13 = 3133$

45. $223 \times 53 = (22 \times 53 \times 10) + (3 \times 53) = 11{,}660 + 159 = 11{,}819$

46. $682 \times 82 = 600 \times 82 + 82^2 = 49{,}200 + 6724 = 55{,}924$

Estimate the following 2-digit cubes.

47. $27^3 \approx 30 \times 30 \times 21 = 30 \times 630 = 18{,}900$

48. $51^3 \approx 50 \times 50 \times 53 = 50 \times 2650 = 132{,}500$

49. $72^3 \approx 70 \times 70 \times 76 = 70 \times 5320 = 372{,}400$

50. $99^3 \approx 100 \times 100 \times 97 = 970{,}000$

51. $66^3 \approx 70 \times 70 \times 58 = 70 \times 4060 = 284{,}200$

BONUS MATERIAL: We can also compute the exact value of a cube with only a little more effort. For example, to cube 42, we use $z = 40$ and $d = 2$. The approximate cube is $40 \times 40 \times 46 = 73{,}600$. To get the exact cube, we can use the following algebra: $(z + d)^3 = z(z(z + 3d) + 3d^2) + d^3$. First, we do $z(z + 3d) + 3d^2 = 40 \times 46 + 12 = 1852$. Then, we multiply this number by z again: $1852 \times 40 = 74{,}080$. Finally, we add $d^3 = 2^3 = 8$ to get 74,088.

Notice that when cubing a 2-digit number, in our first addition step, the value of $3d^2$ can be one of only five numbers: 3, 12, 27, 48, or 75. Specifically, if the number ends in 1 (so $d = 1$) or ends in 9 (so $d = -1$), then $3d^2 = 3$. Similarly, if the last digit is 2 or 8, we add 12; if it's 3 or 7, we add 27; if it's 4 or 6, we add 48; if it's 5, we add 75. Then, in the last step, we will always add or subtract one of five numbers, based on d^3. Here's the pattern:

If last digit is…	1	2	3	4	5	6	7	8	9
Adjust by…	+1	+8	+27	+64	+125	–64	–27	–8	–1

For example, what is the cube of 96? Here, $z = 100$ and $d = -4$. The approximate cube would be $100 \times 100 \times 88 = 880{,}000$. For the exact cube, we first do $100 \times 88 + 48 = 8848$. Then we multiply by 100 and subtract 64: $8848 \times 100 - 64 = 884{,}800 - 64 = 884{,}736$.

Using these examples as a guide, compute the exact values of the following cubes.

52. $13^3 = (10 \times 19 + 27) \times 10 + 3^3 = 2170 + 27 = 2197$

53. $19^3 = (20 \times 17 + 3) \times 20 + (-1)^3 = 343 \times 20 - 1 = 6859$

54. $25^3 = (20 \times 35 + 75) \times 20 + 5^3 = 775 \times 20 + 125 = 15{,}500 + 125 = 15{,}625$

55. $59^3 = (60 \times 57 + 3) \times 60 + (-1)^3 = 3423 \times 60 - 1 = 205{,}379$

(Note: $3423 \times 6 = 3400 \times 6 + 23 \times 6 = 20{,}400 + 138 = 20{,}538$)

56. $72^3 = (70 \times 76 + 12) \times 70 + 2^3 = 5332 \times 70 + 8 = 373{,}248$

(Note: $5332 \times 7 = 5300 \times 7 + 32 \times 7 = 37{,}100 + 224 = 37{,}324$)

Lecture 12

We begin this section with a sample of review problems. Most likely, these problems would have been extremely hard for you to do before this course began, but I hope that now they won't seem so bad.

1. If an item costs \$36.78, how much change would you get from \$100?

Because the dollars sum to 99 and the cents sum to 100, the change is \$63.22.

2. Do the mental subtraction problem: 1618 – 789.

$1618 - 789 = 1618 - (800 - 11) - 818 + 11 = 829$

Do the following multiplication problems.

3. $13 \times 18 = (13 + 8) \times 10 + (3 \times 8) = 210 + 24 = 234$

4. $65 \times 65 = 60 \times 70 + 5^2 = 4200 + 25 = 4225$

5. $997 \times 996 = (1000 \times 993) + (-3) \times (-4) = 993{,}012$

6. Is the number 72,534 a multiple of 11?

Yes, because $7 - 2 + 5 - 3 + 4 = 11$.

7. What is the remainder when you divide 72,534 by a multiple of 9?

Because $7 + 2 + 5 + 3 + 4 = 21$, which sums to 3, the remainder is 3.

8. Determine 23/7 to 6 decimal places.

$23/7 = 3\ 2/7 = 3.285714$ (repeated)

9. If you multiply a 5-digit number beginning with 5 by a 6-digit number beginning with 6, then how many digits will be in the answer?

Just from the number of digits in the problem, you know the answer must be either 11 digits or 10 digits. Then, because the product of the initial digits in this particular problem ($5 \times 6 = 30$), is more than 10, the answer is definitely the longer of the two choices, in this case 11 digits.

10. Estimate the square root of 70.

$70 \div 8 = 8\ 3/4 = 8.75$. Averaging 8 and 8.75 gives us an estimate of 8.37.

(Exact answer begins 8.366… .)

Do the following problems on paper and just write down the answer.

11. $509 \times 325 = 165{,}425$ (by criss-cross method).

12. $21{,}401 \div 9$: Using the Vedic method, we get 2 3 7 7 R 8.

13. $34{,}567 \div 89$: Using the Vedic method, with divisor 9 and multiplier 1, we get:

$$\begin{array}{r|l} & \ \ 3\ 8\ 8 \quad R\,35 \\ \hline 89 & 3\ 4_7 5_6 6_2 7 \\ +1 & \\ \hline 90 & \end{array}$$

14. Use the phonetic code to memorize the following chemical elements: Aluminum is the 13th element, copper is the 29th element, and lead is the 82nd element.

Aluminum = 13 = DIME or TOMB. An aluminum can filled with DIMEs or maybe a TOMBstone that was "Aluminated"?

Copper = 29 = KNOB or NAP. A doorKNOB made of copper or a COP taking a NAP.

Lead = 82 = VAN or FUN. A VAN filled with lead pipes or maybe being "lead" to a FUN event.

15. What day of the week was March 17, 2000? Day $= 2 + 17 + 0 - 14 = 5$ = Friday.

16. Compute $212^2 = 200 \times 224 + 12^2 = 44{,}800 + 144 = 44{,}944$

17. Why must the cube root of a 4-, 5-, or 6-digit number be a 2-digit number?

The largest 1-digit cube is $9^3 = 729$, which has 3 digits, and a 3-digit cube must be at least $100^3 = 1{,}000{,}000$, which has 7 digits.

Find the cube roots of the following numbers.

18. 12,167 has cube root 23.

19. 357,911 has cube root 71.

20. 175,616 has cube root 56.

21. 205,379 has cube root 59.

The next few problems will allow us to find the cube root when the original number is the cube of a 3-digit number. We'll first build up some ideas to find the cube root of 17,173,512, which is the cube of a 3-digit number.

22. Why must the first digit of the answer be 2?

$200^3 = 8{,}000{,}000$ and $300^3 = 27{,}000{,}000$, so the answer must be in the 200s.

23. Why must the last digit of the answer be 8?

Because 8 is the only digit that, when cubed, ends in 2.

24. How can we quickly tell that 17,173,512 is a multiple of 9?

By adding its digits, which sum to 27, a multiple of 9.

25. It follows that the 3-digit number must be a multiple of 3 (because if the 3-digit number was not a multiple of 3, then its cube could not be a multiple of 9). What middle digits would result in the number 2_8 being a multiple of 3? There are three possibilities.

For 2_8 to be a multiple of 3, its digits must sum to a multiple of 3. This works only when the middle number is 2, 5, or 8 because the digit sums of 228, 258, and 288 are 12, 15, and 18, respectively.

26. Use estimation to choose which of the three possibilities is most reasonable.

Since 17,000,000 is nearly halfway between 8,000,000 and 27,000,000, the middle choice, 258, seems most reasonable. Indeed, if we approximate the cube of 26 as $30 \times 30 \times 22 = 19{,}800$, we get 260^3, which is about 20 million, consistent with our answer.

Using the steps above, we can do cube roots of any 3-digit cubes. The first digit can be determined by looking at the millions digits (the numbers before the first comma); the last digit can be determined by looking at the last digit of the cube; the middle digit can be determined through digit sums and estimation. There will always be three or four possibilities for the middle digit; they can be determined using the following observations, which you should verify.

27. Verify that if the digit sum of a number is 3, 6, or 9, then its cube will have digit sum 9.

If the digit sum is 3, 6, or 9, then the number is a multiple of 3, which when cubed will be a multiple of 9; thus, its digits will sum to 9.

28. Verify that if the digit sum of a number is 1, 4, or 7, then its cube will have digit sum 1.

A number with digit sum 1, when cubed, will have a digit sum that can be reduced to $1^3 = 1$. Likewise, $4^3 = 64$ reduces to 1 and $7^3 = 343$ reduces to 1.

29. Verify that if the digit sum of a number is 2, 5, or 8, then its cube will have digit sum 8.

Similarly, a number with digit sum 2, 5, or 8, when cubed, will have the same digit sum as $2^3 = 8$, $5^3 = 125$, and $8^3 = 512$, respectively, all of which have digit sum 8.

Using these ideas, determine the 3-digit number that produces the cubes below.

30. Find the cube root of 212,776,173.

Since $5^3 < 212 < 6^3$, the first digit is 5, and since 7^3 ends in 3, the last digit is 7. Thus, the answer looks like 5_7. The digit sum of 212,776,173 is 36, which is a multiple of 9, so the number 5_7 must be a multiple of 3. Hence, the middle digit must be 0, 3, 6, or 9 (because the digit sums of 507, 537, 567, and 597 are all multiples of 3). Given that 212,000,000 is so close to 600^3 (= 216,000,000), we pick the largest choice: 597.

31. Find the cube root of 374,805,361.

Since $7^3 < 374 < 8^3$, the first digit is 7, and since only 1^3 ends in 1, the last digit is 1. Thus, the answer looks like 7_1. The digit sum of 74,805,361 is 37, which has digit sum 1; by our previous observation, 7_1 must have a digit sum that reduces to 1, 4, or 7. Hence, the middle digit must be 2, 5, or 8 (because 721, 751, and 781 have digit sums 10, 13, and 16, which reduce to 1, 4, and 7, respectively). Given that 374 is much closer to 343 than it is to 512, we choose the smallest possibility, 721. To be on the safe side, we estimate 72^3 as $70 \times 70 \times 76 = 372{,}400$, which means that 720^3 is about 372,000,000; thus, the answer 721 must be correct.

32. Find the cube root of 4,410,944.

Here, $1^3 < 4 < 2^3$, so the first digit is 1, and (by examining the last digit) the last digit must be 4. Hence, the answer looks like 1_4. The digit sum of 4,410,944 is 26, which reduces to 8, so 1_4 must reduce to 2, 5, or 8. Thus, the middle digit must be 0, 3, 6, or 9. Given that 4 is comfortably between 1^3 and 2^3, it must be 134 or 164. Since 163 = 16 × 16 × 16 = 256 × 8 × 2 = 2048 × 2 = 4096, we choose the answer 164.

Compute the following 5-digit squares in your head! (Note that the necessary 2-by-3 and 3-digit square calculations were given in the solutions to Lecture 11.)

33. $11{,}235^2$

11 × 235 × 2 = 2585 × 2 = 5,170. So 11,000 × 235 × 2 = 5,170,000.

We can hold the 5 on our fingers and turn 170 into DUCKS. $11{,}000^2$ = 121,000,000, which when added to 5 million gives us 126 million, which we can say. Next, we have 235^2 = 55,225, which when added to 170,000, gives us the rest of the answer: 225,225. Final answer = 126,225,225.

34. $56{,}753^2$

56,000 × 753 × 2 = 56 × 753 × 2 × 1000 = 42,168 × 2 × 1000 = 84,336,000 = FIRE, MY MATCH.

$56{,}000^2$ = 3,136,000,000, so we can say "3 billion." After adding 136 to 84 (FIRE), we can say "220 million." Then, 753^2 = 567,009, which when added to 336,000 (MY MATCH) gives the rest of the answer, 903,009. Final answer = 3,220,903,009.

35. $82{,}682^2$

$82{,}000 \times 682 \times 2 = 82 \times 682 \times 2 \times 1000 = 55{,}924 \times 2 \times 1000$ $= 111{,}848$ = DOTTED, VERIFY.

$82{,}000^2 = 6{,}724{,}000{,}000$, so we can say "6 billion," then add 724 to 111 (DOTTED) to get 835 million, but because we see a carry coming (from $848{,}000 + 682^2$), we say "836 million." Next, $682^2 = 465{,}124$ (turning 124 into TENOR, if helpful). Now, $465{,}000 + 848{,}000$ (VERIFY) $= 1{,}313{,}000$, but we have already taken care of the leading 1, so we can say "313 thousand," followed by (TENOR) 124. Final answer = 6,836,313,124.

Timeline

B.C.

46 .... Julian calendar established.

A.D.

c. 500 .... Hindu mathematicians originate positional notation for numbers and most techniques of arithmetic using that notation.

c. 900 .... Decimal fractions in use in the Arab world.

1202 .... Publication of *Liber Abaci*, by Leonardo of Pisa (aka Fibonacci). This book introduced Arabic numerals and Hindu techniques of arithmetic to the Western world.

1582 .... Gregorian calendar established. Thursday, October 4, 1582, was followed by Friday, October 15, 1582, for all countries that adopted it at that time.

1634 .... Early phonetic code introduced by the French mathematician Pierre Hérigone (1580–1643).

1699 .... German Protestant states adopt the Gregorian calendar.

1730 .... Phonetic code using both vowels and consonants developed by Rev. Richard Grey.

1752 .... England and the English colonies adopt the Gregorian calendar.

1804 .... Birth of lightning calculator Zerah Colburn.

1806 .... Birth of lightning calculator George Parker Bidder.

1807 .... Phonetic code that assigned only consonant sounds to the digits 0 to 9 developed by Gregor von Feinagle, a German monk.

1820 .... Aimé Paris creates a more user-friendly version of Feinagle's phonetic code, which became the Major system in use today.

1938 .... Physicist Frank Benford states Benford's law: For many types of data, the first digit is most likely to be 1, then 2, then 3, and so on, with 9 the least common first digit of all.

1960 .... *The Trachtenberg Speed System of Basic Mathematics* by Jakow Trachtenberg published in English.

1965 .... Posthumous publication of *Vedic Mathematics* by Bhāratī Krishna Tirthajī.

2004 .... Mental Calculation World Cup first held.

Glossary

addition method: A method for multiplying numbers by breaking the problem into sums of numbers. For example, $4 \times 17 = (4 \times 10) + (4 \times 7) = 40 + 28 = 68$, or $41 \times 17 = (40 \times 17) + (1 \times 17) = 680 + 17 = 697$.

associative law: A law of multiplication that for any numbers a, b, c, $(a \times b) \times c = a \times (b \times c)$. For example, $23 \times 16 = 23 \times (8 \times 2) = (23 \times 8) \times 2$. There is also an associative law of addition: $(a + b) + c = a + (b + c)$.

Benford's law: The phenomenon that most of the numbers we encounter begin with smaller digits rather than larger digits. Specifically, for many real-world problems (from home addresses, to tax returns, to distances to galaxies), the first digit is N with probability $\log(N+1) - \log(N)$, where $\log(N)$ is the base 10 logarithm of N satisfying $10^{\log(N)} = N$.

casting out nines (also known as the method of digit sums): A method of verifying an addition, subtraction, or multiplication problem by reducing each number in the problem to a 1-digit number obtained by adding the digits. For example, 67 sums to 13, which sums to 4, and 83 sums to 11, which sums to 2. When verifying that $67 + 83 = 150$, we see that 150 sums to 6, which is consistent with $4 + 2 = 6$. When verifying $67 \times 83 = 5561$, we see that 5561 sums to 17 which sums to 8, which is consistent with $4 \times 2 = 8$.

close-together method: A method for multiplying two numbers that are close together. When the close-together method is applied to 23×26, we calculate $(20 \times 29) + (3 \times 6) = 580 + 18 = 598$.

complement: The distance between a number and a convenient round number, typically, 100 or 1000. For example, the complement of 43 is 57 since $43 + 57 = 100$.

create-a-zero, kill-a-zero method: A method for testing whether a number is divisible by another number by adding or subtracting a multiple of the second number so that the original number ends in zero.

criss-cross method: A quick method for multiplying numbers on paper. The answer is written from right to left, and nothing else is written down.

cube root: A number that, when cubed, produces a given number. For example, the cube root of 8 is 2 since $2 \times 2 \times 2 = 8$.

cubing: Raising a number to the third power. For example, the cube of 4, denoted 4^3, is equal to 64.

distributive law: The rule of arithmetic that combines addition with multiplication, specifically $a \times (b + c) = (a \times b) + (a \times c)$.

factoring method: A method for multiplying numbers by factoring one of the numbers into smaller parts. For example, $35 \times 14 = 35 \times 2 \times 7 = 70 \times 7 = 490$.

Gregorian calendar: Established by Pope Gregory XIII in 1582, it replaced the Julian calendar to more accurately reflect the length of the Earth's average orbit around the Sun; it did so by allowing three fewer leap years for every 400 years. Under the Julian calendar, every 4 years was a leap year, even when the year was divisible by 100.

leap year: A year with 366 days. According to our Gregorian calendar, a year is usually a leap year if it is divisible by 4. However, if the year is divisible by 100 and not by 400, then it is not a leap year. For example, 1700, 1800, and 1900 are not leap years, but 2000 is a leap year. In the 21st century, 2004, 2008, …, 2096 are leap years, but 2100 is not a leap year.

left to right: The "right" way to do mental math.

Major system: A phonetic code that assigns consonant sounds to digits. For example 1 gets the *t* or *d* sound, 2 gets the *n* sound, and so on. By inserting vowel sounds, numbers can be turned into words, which make them easier to remember. It is named after Major Beniowski, a leading memory expert in London, although the code was developed by Gregor von Feinagle and perfected by Aimé Paris.

math of least resistance: Choosing the easiest mental calculating strategy among several possibilities. For example, to do the problem 43 × 28, it is easier to do 43 × 7 × 4 = 301 × 4 = 1204 than to do 43 × 4 × 7 = 172 × 7.

peg system: A way to remember lists of objects, especially when the items of the list are given a number, such as the list of presidents, elements, or constitutional amendments. Each number is turned into a word using a phonetic code, and that word is linked to the object to be remembered.

right to left: The "wrong" way to do mental math.

square root: A number that, when multiplied by itself, produces a given number. For example, the square root of 9 is 3 and the square root of 2 begins 1.414…. Incidentally, the square root is defined to be greater than or equal to zero, so the square root of 9 is *not* –3, even though –3 multiplied by itself is also 9.

squaring: Multiplying a number by itself. For example, the square of 5 is 25.

subtraction method: A method for multiplying numbers by turning the original problem into a subtraction problem. For example, 9 × 79 = (9 × 80) – (9 × 1) = 720 – 9 = 711, or 19 × 37 = (20 × 37) – (1 × 37) = 740 – 37 = 703.

Vedic mathematics: A collection of arithmetic and algebraic shortcut techniques, especially suitable for pencil and paper calculations, that were popularized by Bhāratī Krishna Tirthajī in the 20th century.

Bibliography

The short list of books:

The books I would most recommend for this course are those by Benjamin and Shermer, Higbee, and Kelly. All three of these paperback books can be found for less than the price of a typical college textbook.

Benjamin, Arthur, and Michael Shermer. *Secrets of Mental Math: The Mathemagician's Guide to Lightning Calculation and Amazing Math Tricks*. New York: Three Rivers Press, 2006. (Also published in the United Kingdom by Souvenir Press Ltd., London, with the title: *Think Like a Maths Genius*. An earlier version of this book was published in 1993 by Contemporary Books in Chicago with the title *Mathemagics: How to Look Like a Genius Without Really Trying*.) This is essentially the book on which this entire course is based. It contains nearly all the topics of this course (except for Vedic division) as well as other amazing feats of mind.

Cutler, Ann, and Rudolph McShane. *The Trachtenberg Speed System of Basic Mathematics*. New York: Doubleday, 1960. This book focuses primarily on problems that involve paper, such as multiplying numbers using the criss-cross method, casting out nines, and adding up long columns of numbers. Everything is done from right to left.

Doerfler, Ronald W. *Dead Reckoning: Calculating Without Instruments*. Houston, TX: Gulf Publishing Co., 1993. An advanced book on doing higher mathematics in your head, going well beyond simple arithmetic. You'll learn how to do (without a calculator, of course) square roots, cube roots, higher roots, logarithms, trigonometric functions, and inverse trigonometric functions.

Duncan, David Ewing. *The Calendar: The 5000-Year Struggle to Align the Clock and the Heavens—and What Happened to the Missing Ten Days*. London: Fourth Estate Ltd., 1998. An enjoyable read about the history of the calendar, from ancient times through the Gregorian calendar.

Flansburg, Scott, and Victoria Hay. *Math Magic: The Human Calculator Shows How to Master Everyday Math Problems in Seconds.* New York: William Morrow and Co., 1993. Focuses primarily on problems suitable for paper (e.g., adding columns of numbers, criss-cross, multiplying numbers close to 100 or 1000, and casting out nines), along with basic information about percentages, decimals, fractions, and such applications as measurements and areas.

Handley, Bill. *Speed Mathematics: Secrets of Lightning Mental Calculation.* Australia: John Wiley and Sons, 2000. Includes some interesting extensions of the close-together method and the calculation of square roots.

Higbee, Kenneth L. *Your Memory: How It Works and How to Improve It.* Cambridge, MA: Da Capo Press, 2001 (1977). Written by a professor of psychology, this book teaches techniques for memorizing names, faces, lists, numbers, and foreign vocabulary. The book includes many references to the medical and psychological literature to gain a deeper appreciation for how mnemonics work.

Hope, Jack A., Barbara J. Reys, and Robert E. Reys. *Mental Math in the Middle Grades*. Palo Alto, CA: Dale Seymour Publications, 1987. See also *Mental Math in Junior High* and *Mental Math in the Primary Grades* by the same authors and publisher. This is a workbook for students in grades 4-6, introducing the fundamentals of left-to-right arithmetic and looking for exploitable features of problems. The other books cover similar topics for grades 7–9 and 1–3, respectively.

Julius, Edward H. *More Rapid Math Tricks and Tips: 30 Days to Number Mastery*. New York: John Wiley and Sons, 1996. This book has much in common with *Rapid Math Tricks and Tips* but has enough new content (especially for division) to make the book worthwhile. Julius has two other books on the market (*Rapid Math in 10 Days* and *Arithmetricks*), but most of the material in these books appears in *Rapid Math Tricks and Tips* and *More Rapid Math Tricks and Tips*.

———. *Rapid Math Tricks and Tips: 30 Days to Number Power.* New York: John Wiley and Sons, 1992. This book has useful suggestions for getting started with mental calculation and sections on the basics of mental addition, subtraction, and multiplication; the criss-cross method; amusing parlor tricks; and special problems (e.g., multiply by 25, divide by 12, square numbers that end in 1, and so on).

Kelly, Gerald W. *Short-Cut Math*. New York: Dover Publications, 1984 (1969). A solid overall reference, with good ideas for mental (and paper) mathematics, focusing on addition, subtraction, multiplication, division, estimation, and fractions. Because it's published by Dover, it's very inexpensive.

Lane, George. *Mind Games: Amazing Mental Arithmetic Tricks Made Easy*. London: Metro Publishing, 2004. A world-champion lightning calculator reveals some of his tricks of the trade. The book is written in a somewhat quirky style and is pretty heavy lifting, but it may be of value to someone who wants to compute square roots and higher roots for the Mental Math World Cup.

Lorayne, Harry, and Jerry Lucas. *The Memory Book: The Classic Guide to Improving Your Memory at Work, at School, and at Play*. New York: Ballantine Books, 1996 (1974). This is the book that taught me the phonetic code and other fundamental techniques for memory improvement. Written in a clear and enjoyable style.

Reingold, Edward M., and Nachum Dershowitz. *Calendrical Calculations: The Millennium Edition*. New York: Cambridge University Press, 2001. Provides complete descriptions of virtually every calendrical system ever used (e.g., Gregorian, Julian, Mayan, Hebrew, Islamic, Chinese, Ecclesiastical), along with algorithms to determine days of the week and major holidays. Comes with a CD with implementations of these algorithms, allowing the user to convert dates from one calendar to another.

Rusczyk, Richard. *Introduction to Algebra*. Alpine, CA: AoPS Incorporated, 2009. A great introduction to algebra, published by the Art of Problem Solving (www.ArtOfProblemSolving.com), publisher of math books for smart people. Covers all topics in Algebra I and some topics in Algebra II.

AoPS also has terrific books on intermediate algebra, geometry, number theory, probability and counting, problem solving, pre-calculus, and calculus.

Ryan, Mark. *Everyday Math for Everyday Life: A Handbook for When It Just Doesn't Add Up*. If you are so rusty with your math skills that you want to start from scratch, this would be a good book to use. The book focuses on hand calculation and mental estimation skills, along with real-life applications of math, such as measurements, checkbook tips, and unit conversions.

Smith, Steven B. *The Great Mental Calculators: The Psychology, Methods and Lives of Calculating Prodigies Past and Present*. New York: Columbia University Press, 1983. The title says it all. Professor Arthur Benjamin is the only living American profiled in this book.

Tekriwal, Gaurav. *5 DVD Set on Vedic Maths*. www.vedicmathsindia.org/dvd.htm, 2009. Provides video instruction on Vedic mathematics, taught by the president of the Vedic Maths Forum in India. The instructor goes through 10 hours worth of problems, standing in front of a whiteboard. Among the topics included are the close-together method, the criss-cross method, Vedic division, and solutions of various algebraic equations.

Tirthajī, Bhāratī Krishna. *Vedic Mathematics*. Delhi: Motilal Banarsidass Publishers Private Ltd., 1992 (1965). The book from which all other books on Vedic mathematics are derived. A good deal of material is presented on mental arithmetic (mostly for pencil-and-paper purposes) and algebra, including much that is not covered in this course. The book is somewhat challenging to read because of the quality of exposition and some of its notation.

Weinstein, Lawrence, and John Adam. *Guesstimation: Solving the World's Problems on the Back of a Cocktail Napkin*. Princeton, NJ: Princeton University Press, 2008. Written by two physicists using nothing more than basic arithmetic, this book provides interesting strategies for coming up with reasonable estimates (within a factor of 10) of problems that initially sound impossible to comprehend. Filled with plenty of interesting examples, such as how many golf balls would be needed to circle the equator or how many acres of farmland would be required to fuel your car with ethanol.

Williams, Kenneth, and Mark Gaskell. *The Cosmic Calculator: A Vedic Mathematics Course for Schools, Book 3*. New Delhi: Motilal Banarsidass Publishers, 2002. This book describes, using notation different from mine, the Vedic method for division problems and an interesting method for doing square roots, along with topics from algebra, geometry, and probability that do not pertain to mental calculation. Two other books with the same name are also available that cover similar topics.

Other books on related topics:

Burns, Marilyn. *Math for Smarty Pants*. Illustrated by Martha Weston. Boston: Little, Brown, and Co., 1982. This is the best book on this list that is aimed at kids. Lots of fun material, illustrated with great cartoons. Filled with mathematical magic tricks, number puzzles, calculation tricks, and paradoxes. If you like this book, then you should also get *The I Hate Mathematics! Book* by the same author and illustrator.

Butterworth, Brian. *What Counts: How Every Brain Is Hardwired for Math*. New York: Free Press, 1999. An interesting book on how the mind represents mathematics and how the brain has developed to count, do arithmetic, and reason about mathematics.

Dehaene, Stanislas. *The Number Sense: How the Mind Creates Mathematics*. New York: Oxford University Press, 1997. A fascinating account of how animals and humans (including babies, autistic savants, and calculating prodigies) conceptualize numbers.

Gardner, Martin. *Aha! Insight!* and *Aha! Gotcha!* Washington, DC: Mathematical Association of America, 2006. Gardner has written dozens of books on recreational mathematics and turned on more people to mathematics than anyone else. These two books are sold as one and contain ingenious mathematical puzzles for which the best solutions require you to think outside the box. Suitable for children and adults.

Lorayne, Harry. *How to Perform Feats of Mathematical Wizardry*. New York: Harry Lorayne, 2006. This book is written for magicians who wish to amaze their audiences with amazing feats of mind and other mathematically based tricks.

Sticker, Henry. *How to Calculate Quickly*. New York: Dover Publications, 1955. A collection of 383 groups of problems (literally, more than 9000 problems) designed to give you practice at doing mental arithmetic. It's mostly problems without a lot of exposition. If you are looking for an inexpensive Dover book, the book by Kelly is superior.

Stoddard, Edward. *Speed Mathematics Simplified*. New York: Dover Publications, 1994 (1965). This book takes a radically different approach from all the other books and is motivated by the system for using a manual abacus. For example, to add 8 to a number, subtract 2, then add 10. This idea eliminates the need for nearly half of the addition table and shows new ways to represent addition, subtraction, multiplication, and division problems, all done from left to right. It's an interesting approach that some might appreciate, but the methods taught in the Stoddard book are very different from the ones taught in this course.

Internet resources:

Art of Problem Solving. www.ArtOfProblemSolving.com. Publisher of outstanding mathematics books (from algebra to calculus) aimed at high-ability students and adults, AoPS also offers online classes and an online community for students, parents, and teachers to share ideas.

Doerfler, Ronald. www.myreckonings.com/wordpress/. Lost arts in the mathematical sciences, including several interesting pages about the history and techniques of lightning calculators.

Mathematical Association of America. www.maa.org. The premier organization in the United States dedicated to the effective communication of mathematics. Publisher of hundreds of interesting mathematics books, particularly at the college level.

Memoriad. www.memoriad.com. The Web site for the World Mental Calculation, Memory and Photographic Reading Olympiad.

Phonetic Mnemonic Major Memory System. http://www.phoneticmnemonic.com/. A dictionary that has converted more than 13,000 words into numbers using the phonetic code. Free and easy to use.

Notes